OBJECTLESSONS

A book series about the hidden lives of ordinary things.

Series Editors:

Ian Bogost and Christopher Schaberg

W0246598

In association with

Program
in Public Scholarship

Washington
University in St.Louis

The Object Lessons series achieves something very close to magic: the books take ordinary—even banal—objects and animate them with a rich history of invention, political struggle, science, and popular mythology. Filled with fascinating details and conveyed in sharp, accessible prose, the books make the everyday world come to life. Be warned: once you've read a few of these, you'll start walking around your house, picking up random objects, and musing aloud: 'I wonder what the story is behind this thing?'"

Steven Johnson, author of *Where Good Ideas Come From* and *How We Got to Now*

Object Lessons describes themselves as 'short, beautiful books,' and to that, I'll say, amen. . . . If you read enough Object Lessons books, you'll fill your head with plenty of trivia to amaze and annoy your friends and loved ones—caution recommended on pontificating on the objects surrounding you. More importantly, though . . . they inspire us to take a second look at parts of the everyday that we've taken for granted. These are not so much lessons about the objects themselves, but opportunities for self-reflection and storytelling. They remind us that we are surrounded by a wondrous world, as long as we care to look."

John Warner, *The Chicago Tribune*

"Besides being beautiful little hand-sized objects themselves, showcasing exceptional writing, the wonder of these books is that they exist at all . . . Uniformly excellent, engaging, thought-provoking, and informative."

Jennifer Bort Yacovissi, *Washington Independent Review of Books*

". . . edifying and entertaining . . . perfect for slipping in a pocket and pulling out when life is on hold."

Sarah Murdoch, *Toronto Star*

"For my money, Object Lessons is the most consistently interesting nonfiction book series in America."

PopMatters

"Though short, at roughly 25,000 words apiece, these books are anything but slight."

Marina Benjamin, *New Statesman*

"[W]itty, thought-provoking, and poetic . . . These little books are a page-flipper's dream."

John Timpane, *The Philadelphia Inquirer*

"The joy of the series, of reading *Remote Control*, *Golf Ball*, *Driver's License*, *Drone*, *Silence*, *Glass*, *Refrigerator*, *Hotel*, and *Waste* (more titles are listed as forthcoming) in quick succession, lies in encountering the various turns through which each of their authors has been put by his or her object. As for Benjamin, so for the authors of the series, the object predominates, sits squarely center stage, directs the action. The object decides the genre, the chronology, and the limits of the study. Accordingly, the author has to take her cue from the *thing* she chose or that chose her. The result is a wonderfully uneven series of books, each one a *thing* unto itself."

Julian Yates, *Los Angeles Review of Books*

"The Object Lessons series has a beautifully simple premise. Each book or essay centers on a specific object. This can be mundane or unexpected, humorous or politically timely. Whatever the subject, these descriptions reveal the rich worlds hidden under the surface of things."

Christine Ro, *Book Riot*

". . . a sensibility somewhere between Roland Barthes and Wes Anderson."

Simon Reynolds, author of *Retromania: Pop Culture's Addiction to Its Own Past*

"My favourite series of short pop culture books"

Zoomer magazine

BOOKS IN THE SERIES

Air Conditioning by Hsuan L. Hsu
Alarm by Alice Bennett
Barcode by Jordan Frith
Bicycle by Jonathan Maskit
Bird by Erik Anderson
Blackface by Ayanna Thompson
Blanket by Kara Thompson
Blue Jeans by Carolyn Purnell
Bookshelf by Lydia Pyne
Bread by Scott Cutler Shershow
Bulletproof Vest by Kenneth R. Rosen
Burger by Carol J. Adams
Cell Tower by Steven E. Jones
Cigarette Lighter by Jack Pendarvis
Coffee by Dinah Lenney
Compact Disc by Robert Barry
Doctor by Andrew Bomback
Dust by Michael Marder
Earth by Jeffrey Jerome Cohen and Linda T. Elkins-Tanton
Egg by Nicole Walker
Email by Randy Malamud
Environment by Rolf Halden
Exit by Laura Waddell
Eye Chart by William Germano
Fat by Hanne Blank
Glass by John Garrison
Golf Ball by Harry Brown
Doll Maria Teresa Hart
Driver's License by Meredith Castile
Drone by Adam Rothstein
Fake by Kati Stevens
Football by Mark Yakich
Gin by Shonna Milliken Humphrey
Glitter by Nicole Seymour
Grave Allison C. Meier
Hair by Scott Lowe
Hashtag by Elizabeth Losh
High Heel by Summer Brennan
Hood by Alison Kinney
Hotel by Joanna Walsh
Hyphen by Pardis Mahdavi
Island by Julian Hanna
Jet Lag by Christopher J. Lee
Luggage by Susan Harlan
Magazine by Jeff Jarvis
Magnet by Eva Barbarossa
Mask by Sharrona Pearl
Mushroom by Sara Rich
Newspaper by Maggie Messitt
Ocean by Steve Mentz
Office by Sheila Liming
OK by Michelle McSweeney
Password by Martin Paul Eve
Pencil by Carol Beggy
Perfume by Megan Volpert
Personal Stereo by Rebecca Tuhus-Dubrow
Phone Booth by Ariana Kelly
Pill by Robert Bennett
Political Sign by Tobias Carroll
Potato by Rebecca Earle
Pregnancy Test by Karen Weingarten
Questionnaire by Evan Kindley
Recipe by Lynn Z. Bloom
Refrigerator by Jonathan Rees
Remote Control by Caetlin Benson-Allott
Rust by Jean-Michel Rabaté
Sewer by Jessica Leigh Hester
Scream by Michael J. Seidlinger
Shipping Container by Craig Martin
Shopping Mall by Matthew Newton
Signature by Hunter Dukes
Silence by John Biguenet
Skateboard by Jonathan Russell Clark
Sock by Kim Adrian
Souvenir by Rolf Potts
Snake by Erica Wright
Spacecraft by Timothy Morton
Space Rover by Stewart Lawrence Sinclair
Sticker by Henry Hoke
Stroller by Amanda Parrish Morgan
Swimming Pool by Hsuan L. Hsu
Traffic by Paul Josephson
Tree by Matthew Battles
Trench Coat by Jane Tynan
Tumor by Anna Leahy
TV by Susan Bordo
Veil by Rafia Zakaria
Waste by Brian Thill
Whale Song by Margret Grebowicz
Wine by Meg Bernhard
X-ray by Nicole Lobdell

Island

JULIAN HANNA

BLOOMSBURY ACADEMIC
NEW YORK • LONDON • OXFORD • NEW DELHI • SYDNEY

BLOOMSBURY ACADEMIC
Bloomsbury Publishing Inc
1385 Broadway, New York, NY 10018, USA
50 Bedford Square, London, WC1B 3DP, UK
29 Earlsfort Terrace, Dublin 2, Ireland

BLOOMSBURY, BLOOMSBURY ACADEMIC and the Diana logo are trademarks
of Bloomsbury Publishing Plc

First published in the United States of America 2024

Cover design: Alice Marwick

For legal purposes the Acknowledgments on p. 165 constitute an extension of this copyright page.

Bloomsbury Publishing Inc does not have any control over, or responsibility for, any third-party
websites referred to or in this book. All internet addresses given in this book were correct at the
time of going to press. The author and publisher regret any inconvenience caused if addresses
have changed or sites have ceased to exist, but can accept no responsibility for any such changes.

Whilst every effort has been made to locate copyright holders the publishers would be grateful to
hear from any person(s) not here acknowledged.

Library of Congress Cataloging-in-Publication Data
Names: Hanna, Julian, author.
Title: Island / Julian Hanna.
Description: New York, NY : Bloomsbury Academic, 2024. | Series: Object lessons |
Includes bibliographical references and index.
Identifiers: LCCN 2023058276 (print) | LCCN 2023058277 (ebook) | ISBN 9798765102367
(paperback) | ISBN 9798765102374 (ebook) | ISBN 9798765102381 (pdf)
Subjects: LCSH: Islands. | Hanna, Julian–Travel. | Islands–Miscellanea. | Islands in literature.
Classification: LCC G500 .H26 2024 (print) | LCC G500 (ebook) | DDC 909/.0942–dc23/
eng/20240412
LC record available at https://lccn.loc.gov/2023058276
LC ebook record available at https://lccn.loc.gov/2023058277

ISBN: PB: 979-8-7651-0236-7
ePDF: 979-8-7651-0238-1
eBook: 979-8-7651-0237-4

Series: Object Lessons

Typeset by Deanta Global Publishing Services, Chennai, India
Printed and bound in Great Britain.

To find out more about our authors and books visit www.bloomsbury.com
and sign up for our newsletters.

for Simone, Clyde, and Nico

CONTENTS

1 INTRODUCTION

ISLAND FEVER

"There are no more islands."
—ALBERT CAMUS, "THE MINOTAUR, OR THE STOP IN ORAN"

"There are still blissful islands!"
—FRIEDRICH NIETZSCHE, *THUS SPOKE ZARATHUSTRA*

I'm sitting outside Café Brecht in Amsterdam, watching some men pull old bicycles out of the canal across from the Heineken brewery. Technically, or at least arguably, I'm on an island; or maybe I'm just consoling myself. It is true that this country, much of it reclaimed from the sea, feels more archipelagic than continental, and that Amsterdam has its share of island-like problems, including being a notorious magnet for stag and hen parties (it just launched a tourism campaign with the slogan "Stay Away"[1]). As the capital of the most globalized country in the world, Amsterdam also feels like an island of exiles. *Brave New World* (1932) by Aldous Huxley, whose title derives from Shakespeare's *The Tempest*, also based around exile, and *A Modern Utopia* (1905) by H. G. Wells both feature islands of exiles. Huxley romanticizes those who are banished from his dystopia, whereas Wells sends disruptive counterrevolutionaries to an island to *save* his utopia. Both use islands as prisons, but the key point is that islands are different from mainlands, whether they provide sanctuary or quarantine from mainstream society.

FIGURE 1 Porto Santo, July 2020. All images by the author.

I grew up on an island, and I have lived on several islands as an adult. I've masqueraded at times as a researcher in the field of Island Studies. I was born into a family of islanders: my mother's side from Iceland and Scotland and my father's side from Ireland and Scotland, with Scottish islanders on both sides (Inner and Outer Hebrides). I have two uncles who are as drawn to islands as I am: my mother's younger brother, recently retired and settled again on Vancouver Island (or so we thought) after living in New Zealand, Australia, and Fiji, just slipped the net for a teaching position in Turks and Caicos. Her older brother spent much of his civilian career after the air force as a pilot stationed on various islands, including Mauritius and Taiwan, before moving to Hong Kong to work as a trainer. My mother moved from the middle of Canada to Vancouver Island in the 1970s, to raise me on the southern tip of that idyllic island. Although they split up, my father soon followed: he lived there for twenty years, always within earshot of the sea and the sea lions in the bay, until his early death.

Still, it is hard to declare oneself a true island expert given the sheer number of islands on Earth. That number is difficult to calculate because it depends on counting methods and other shifting criteria; many islands are too small to count, or they come and go with the water level. Norway alone has almost a quarter of a million islands. The global number is likely around a million, but of these only a small fraction are inhabited, perhaps 15,000 or 20,000. Like the guy with the hammer who sees only nails, when you start thinking about islands you realize that everything is potentially an island. Is an automobile on a highway a sort of floating island? Is a duck an island? Is Berghain, the famous nightclub in Berlin with the strict door policy and transcendent dance floor, something like Thomas More's island utopia? And what about digital islands? For the sake of brevity, and sanity, I've decided to stick mainly to a geographic definition with occasional flights of metaphor.

Writers and artists have long sought out islands as places of escape where you can get some work done. (I am editing this introduction on an island.) This was famously the case for George Orwell, who completed *1984* on the Scottish island of Jura; for Tove Jansson, author of the Moomin children's books, who spent thirty summers on the tiny islet of Klovaharun in the Finnish archipelago; and for Ernest Hemingway, living and writing in Cuba and Key West with his polydactyl (six-toed) cats.[2] Ingmar Bergman wrote and set many of his films, including *Persona* (1966), on the Swedish island of Fårö in the Baltic Sea. In the 1920s Robert Graves and Laura Riding said goodbye to all that and escaped the noise of the interwar period for the Spanish island of Majorca. The philosopher Walter Benjamin and the Dada artist Raoul Hausmann were part of the first wave of intellectuals and exiles to hit neighboring Ibiza in the 1930s (as well as the young Albert Camus, whose mother was of Balearic ancestry); the Irish playwright Brendan Behan joined an even larger wave in the 1950s; and after the wave broke in the 1960s with the Rolling Stones and other guests the island was inundated. (Even the depressive Romanian philosopher Emil Cioran found what he was looking for in Ibiza.) Leonard Cohen nurtured his early talent among a small bohemian community on the Greek island of Hydra. The Portuguese Nobel laureate José Saramago lived in self-exile on Lanzarote in the Canary Islands for the last eighteen years of his life. Michel Houellebecq, born on the French colonial island of Réunion, did Lanzarote no favors when he chose it to depict the excesses of holiday island hedonism in his novella of the same name; after the backlash caused by his next novel, *Platform* (2001), he moved to Ireland to live in seclusion. Books, plays, films, and television, from Thomas More's *Utopia* (1516) and *The Tempest* (1611) to *Treasure Island* (1883) and *Peter Pan* (1904) to the reality dating series *Love Island* (2015–), are set on islands because it usefully constrains the action, focusing attention and narrowing the scope of possibilities. *The Tempest* dramatizes this very act: Prospero uses his artistry and magic to orchestrate events on the island he controls.

The very idea of islands in the West is deeply rooted in the Age of Empire. With a nod to Homer's *Odyssey*, as well as *The Tempest* (often staged to highlight the colonial master-slave dynamic of Prospero and the native Caliban), it begins in the modern era with Daniel Defoe's *Robinson Crusoe* (1719), arguably the first English novel. *Crusoe* produced countless offspring, and it encapsulated the idea that a white man can wash up wherever he pleases and claim that land as his own, extract all its valuable commodities, reshape the landscape in the image of his homeland, and kill, convert, or enslave anyone he meets. Jonathan Swift, who had witnessed firsthand in Ireland the destruction colonialism brings to islands, set his savage satire loose on this tendency in *Gulliver's Travels* (1726), in which a Homer Simpson-like character blunders across the ocean, arrogant and clueless, and unfailingly draws the wrong lessons from his experiences in relation to the follies of his own little island of Britain. Out of *Crusoe* came *The Swiss Family Robinson* (1812), *Lord of the Flies* (1954), and J. G. Ballard's *Concrete Island* (1974), in which modern city dwellers live technologically isolated, insular lives. Paul Gauguin's paintings of Tahiti embody the colonial view of an island Eden, primitive and unspoiled (until the ageing artist spread syphilis among the young girls like Teha'amana that he made his "wives"). "Oh, what a secret island," wrote Enid Blyton in her children's adventure book *The Secret Island* (1938), "all for their very own, to live on and play on." The dark side of romantic solitude and adventure, however, is the tendency towards the domination and exploitation of fellow humans and of nature. This colonial mentality lives on in island holiday resorts with advertisements serving up a stereotypical fantasy of tamed, domesticated adventure and luxurious, princely isolation for paying guests from the wealthy mainlands of the Global North—who too often act like modern-day Gullivers, ignorantly pissing on everything and everyone, just as Swift's character did (literally) in Lilliput.

Globalization works both ways, of course, and politicians on the Right in Britain, Australia, and elsewhere have sought to portray their nations as islands under siege. Islands have played an outsized role in migration dramas in recent decades, from Lesbos in Greece and Lampedusa in Italy to the Pacific island of Nauru (which Australia uses for "offshore processing" of refugees) and Diego Garcia, a British territory and militarized atoll in the middle of the Indian Ocean, where dozens of refugees remain stuck in limbo since being stranded in 2021. Islands are frequently idealized as little utopias; Margaritavilles (like Jimmy Buffet's vision of Key West) that offer escape from the pressures of everyday life. Just as often, however, islands are portrayed—and used—as prisons (or jails): Alcatraz, Rikers, Terminal Island, McNeil Island (the prison island in the Pacific Northwest where Charles Manson spent much of the 1960s), Manus Island, Robben Island, Saint Helena, Guantanamo Bay. Sometimes the same island doubles as both prison and paradise. Islands can offer refuge, as they have done to sailors and migrants for centuries, but they have also long been used as offshore prisons, out of sight and mind. As Judith Schalansky notes, remote islands "are well suited as places in which to gather everything that is undesirable, displaced and digressive."[3] How does the paradise island relate to or reconcile with its opposite?

The meaning and definition of islands, their geography and the role they play in the world, is constantly changing. In a sense this has always been the case: since moving to the Netherlands, for example, my wife Simone has become mildly obsessed with Doggerland, the landmass that once connected Britain to mainland Europe, and whose treasures are still scattered on the beaches. But now the internet is making islands less isolated—even as we all may feel increasingly like islanders, living in atomized online communities—and this is rapidly changing their basic character. At the same time, climate change threatens to erase the very existence of many islands. The South Pacific island of Tuvalu, for example, is in a dire situation: it made headlines when the foreign minister delivered a dramatic speech to the Cop26 summit while standing in knee-deep water; it is now working on the creation of a "digital twin" to protect the nation from total extinction when its physical landmass disappears into the sea. The poet and activist Teresia Kieuea Teaiwa of Banaba, a tiny coral island east of Nauru in the Pacific, urges us in her poetry to use island as a verb, to radically alter our outlook and behavior: "Let us 'island' the world! Let us teach the inhabitants of planet Earth how to behave as if we were all living on islands!"[4] As Tove Jansson describes the microcosms of her tiny island in *The Summer Book* (1972), in one part (the magic forest) "the balance between survival and extinction was so delicate that even the smallest change was unthinkable," while the rest of the island was "tidied . . . down to the smallest twig."[5] The highly attuned ways that islanders tend to their islands out of necessity can teach us valuable lessons about mainlands as well.

Island and mainland are always relative terms. I grew up taking the ferry back and forth from Vancouver, on an actual mainland, to an island that was also called Vancouver. The island was a sort of mainland to a chain of smaller islands. In Ireland I heard British people refer without reflection or irony to "the mainland," meaning their own slightly larger island. (The famously self-regarding headline in *The Times* on October 22, 1957, read: "Heavy Fog in the Channel—Continent cut off.") In the Madeiran archipelago, the inhabitants of the second most populous island, Porto Santo, sometimes refer to the main island as the mainland. (They also view the continental mainland as Europe, although the islands lie off the coast of Africa: a reminder that geography is political.) I noticed the same thing when I visited the Aran Islands off the west coast of Ireland, and the Orkney Islands off the north coast of Scotland: the "mainland" was an island, even an island off another island. New Yorkers might not think of a journey from Manhattan to Brooklyn as an island-to-island trip, although technically both boroughs are on islands; yet going from Manhattan out to Fire Island, the barrier islands connected to Long Island by a bridge, or even Montauk at the far end of Long Island, would probably be considered an island getaway—with the feeling of seaside solitude, and the transition from work and responsibilities to the carefree pleasure that islands typically offer. (New York City is rich in island culture: the dozens of islands in its jurisdiction include Liberty and Ellis Island, those two symbols of American hope and promise; Roosevelt Island, with its 1970s utopian housing experiment; Swinburne Island, used for quarantining new arrivals; and Rikers Island, the city's largest jail. The neurologist Oliver Sacks used to swim around City Island in the Bronx, where he lived. Many of the smaller islands dotted around the city have colorful names: Chimney Sweeps Islands, Rat Island, Mau Mau Island, Subway Island.)

One of the epigraphs to this chapter is something Camus wrote that I recently rediscovered: "There are no more islands. Yet there is a need for them. In order to understand the world, one has to turn away from it on occasion." Turning away is how most people end up on islands. That was how my extended family all found themselves on the island where I grew up, far from where any of them were born, and that was how my own family found ourselves hundreds of miles off the coast of Morocco. In many ways the quote sums up the reason for this book, and the paradox at its heart: islands (isolation, insularity) no longer exist as they once did, yet we need them more than ever. But why? The Australian writer Charmian Clift, who described the bohemian milieu on the island of Hydra in the 1950s in her book *Peel Me a Lotus* (1959), describes how on islands "we are stripped to our bare selves."[6] When you are surrounded by the sea it is hard to be anywhere else: the hypnotic sound of breaking waves keeps you present, saying: be here, be here. The writer Peter Conrad, who grew up in Tasmania, insists that "On an island, you are alone, even if you share the place with others. The location is by definition *eccentric*, because it acknowledges that there is a centre elsewhere."[7] Camus's words about "no more islands" were figurative, of course, but they were also true: we are rapidly losing sacred spaces to what David Foster Wallace called "Total Noise."

Remote islands in particular are undergoing huge changes in the age of instant connectivity. Some islands have rebranded themselves as convenient hubs that facilitate "digital nomads," while others (like the institute where I worked in Madeira) pitch themselves as "living labs" for experiments that can be scaled to mainland proportions. Advanced mapping techniques have increased our knowledge of what was once mysterious, changing our view of islands completely. (When Madeira was first spotted from the sea, the sailors refused to land on its imposing shores.) How will islands change further as connectivity increases, and what will be lost? Can islands be both isolated and connected, nodes in a vast network, fragments of land cut off by the sea and yet bound to the mainland by invisible threads (and tangible undersea infrastructure)? In her novel *Flights* (2007), Olga Tokarczuk describes an island "where time . . . turns around disappointed and heads toward land."[8] "Island time" is still a thing that persists—perhaps because we want it to in our increasingly time-obsessed world.

When you visit an island paradise your holiday is in someone else's home. This creates conflict and resentment, even if you don't notice it at first. To some extent in Madeira we took on the local view, rolling our eyes at tourists as we walked to work and ate at our local spots—but we were also always caught between, remaining tourists to the locals and in some sense even to ourselves. As a taxi driver once told me: "There are tourists who visit, and tourists who live here." The statement stuck with me for its shocking frankness; it was a hard lesson to forget. We were trying to escape the rat race, to live a more authentic life, to pry ourselves away from the tight grip of global capitalism—to find, in the end, a more wholesome life (as well as just trying to get by)—and to some degree we found it. But as Madeirans are fond of saying: *é complicado*.

I am obsessed with the BBC radio series Desert Island Discs, which has been running since 1942. My favorite guest is the artist David Hockney, who visited in 1972. Aside from going through a list of eight songs, the interviewer always asks probing questions about personal tragedy and loss, almost like a last confession before the guest is cast away into permanent isolation. Often the guest is brought to tears, but Hockney was arch and cool, choosing songs like "I'm Through with Love." You are also allowed to choose a book. Asked which book he would bring to the island, Hockney chose Floyd Carter's *Route 69* (1968), a book of gay pornography. This was a shocking choice in 1972, but Hockney gave the sensible explanation that anything other than porn would become tedious after you'd read it a few times, and he expected to be lonely, and horny, although he didn't put it in those words. His mother was said to have been bemused by the choice. Recently tour operators have started offering desert island survival holidays in the style of

TV shows like *Naked and Afraid*, where jaded urbanites pay top dollar to be stranded on a desert island armed only with a machete (no Bible or *Collected Works of Shakespeare*, even).[9] A paradise can so easily become a prison, even if the prison is a pleasant and timeless desert island.

Hockney admitted that he would probably go mad if he was confined to a small island. In D. H. Lawrence's short story "The Man Who Loved Islands" (1927), the titular character sets out to make "a world of his own," as the master of a small island. Unlike Hockney, Lawrence fantasized about this kind of escape. But he also saw the perils: soon enough, driven by his growing disenchantment with human society in any form, the man moves to an even smaller island. Before long he finds that even this island feels like a "suburb" and he moves further out to settle on a third, almost uninhabitable island: "a few acres of rock away in the north." What started as a hopeful utopia ends badly, and he is last seen on his desolate rock, thinking feverishly: "The elements! The elements!"[10]

Islands like Madeira, accessible only by boat or plane (weather permitting), can be extremely claustrophobic places, breeding grounds for "island fever." The mysterious ailment is often brought on by the realization that you want to, indeed must, leave an island immediately—even if your visit started as a dream vacation—but you are unable to do so by conventional means. I remember the near-tragic story of an English woman who tried to swim out to a cruise ship as it left Madeira and was found floating on her handbag in the darkness, half a mile offshore.[11] Island fever is a trick of the mind, a classic psychological set up: the more you want to escape, the more you think about it, the more you realize you're trapped, the worse it gets—until you're driven to desperate measures.

This book is built on a few simple constraints (happy ones, I hope): it must be short, it must be some kind of object lesson—a practical illustration of an abstract concept—and it must be about islands. I've tried to gather a taxonomy of island types and lessons: the island as physical and conceptual object; islands as objects for dreams, projections, musings; islands as places to find oneself, lose oneself, find one's dreams, make one's dreams come true, find paradise, make paradise, for pleasure, for punishment. In a more concrete sense, the value of island thinking; how islands can serve as experimental test beds and early warning signs; how they can teach us to adapt, rather than dominating and exploiting; how they bring us closer to others, facilitating chance encounters and utopian communities, as well as closer to ourselves; how islands can help us to see things differently, from the periphery, outside of mainland time and space; and how each island is exceptional, "a little world within itself,"[12] as Darwin wrote of Galapagos with its unique tortoises and finches—later echoed by F. Scott Fitzgerald in his description of West Egg as "a world complete in itself."[13] To be clear, however, this will not be an Island Studies book or a practical guide to islands full of facts and figures. The "lesson" will draw on my own experiences with islands as places of grief, happiness, escape, hope, renewal, and so on, with occasional dips into literary and other accounts. As a guiding motif, each chapter will be a message in a bottle, and each bottle will be specific to that island. The chronology of events will move backwards in time. There are too many examples to work into the narrative, just as there are too many islands to visit in one lifetime. Once you start writing a book, people naturally try to help. "Did you mention Craggy Island?" Simone asked the other day, referring to the cult Irish TV show *Father Ted*, which chronicles the antics of three very odd priests, cast into exile on a remote island with their tea-pushing housekeeper. (No, I had not.) "What about *Gilligan's Island*?" (Yes, somewhere.)

To give the fragmented style of this book a bit more structure, and in keeping with the larger theme, each island tale will be written into a larger archipelago of meaning. Barthes is an obvious inspiration, for example in *Mythologies* (1957) or *Roland Barthes by Roland Barthes* (1975): "To write by fragments: the fragments are then so many stones on the perimeter of a circle . . . at the center, what?"[14] In an early essay titled "En Grèce" (1944), he used the Greek islands themselves as a metaphor for style: "there are so many islands that one does not know if each one is at the center or the edge of an archipelago." Barthes scholar Claude Coste links this passage to the Martinique-born French Caribbean writer Édouard Glissant's concept of "archipelagic thinking," which to my mind means decentralized, rhizomatic, relational, and unpredictable.[15] In John Donne's "Meditation XVII," itself a fragment of *Devotions upon Emergent Occasions* (1623), he claims that: "No man is an island," because each of us is "a piece of the continent, a part of the main." It is a worthy sentiment, and even one that is in tune with climate change: each of us is a humble clod of earth, but when we are washed away by the sea we are missed, as we add an essential part to the whole. Yet perhaps the archipelagic analogy is even more apt: we are all islands, isolated yet connected in a global archipelago. Sometimes a bit of autonomy, after all, in harmony with others, can be a good thing.

A fragmented style will be necessary in any case because I have a lot to say. Living on a remote island will do that to you; it's all bottled up. The last few years have been challenging with the pandemic and its aftermath, and the age of the "polycrisis" now upon us. Islands to me represent both hope and loss. I'm finishing this message in the cafe in Amsterdam: it will go into a Grolsch bottle, bright green with a white porcelain stopper. I suppose it's a cry for help, an SOS, or even a manifesto: *hear my words*. In adventure stories the message in a bottle is a desperate appeal from a lonely, shipwrecked castaway, a futile gesture of isolation, a literal drop in the ocean. But throwing a message into the sea is also an act of hope, however faint. It says: I'm still here, and I'm thinking of you. I will carry the bottles in a wooden crate down to the sea off western Crete, where I plan to finish the book, and drop them into the Mediterranean. I hope that these bottles, wherever they drift, wash up on calmer shores. When the bottle is opened and the scroll of ribbon-tied paper falls into your hand, I hope you catch a faint whiff of the sea and are transported to some far-off island like the ones I will attempt to describe faithfully in the words you are about to read.

2 MOURNING

VANCOUVER ISLAND, SUMMER 2021

Canada had only reopened to international travelers the previous day, and there were still very few of us. I had not seen my mother in three years, but I was here only briefly, and not primarily for a visit but for a funeral service. The last leg of the trip from Amsterdam was the ferry crossing from Tsawwassen to Swartz Bay.

In *City of Glass* (2000), Douglas Coupland describes the ferries from Vancouver to the islands as "wonderfully democratic . . . to walk the deck of one is to garner a good cross-sectional view of society: bureaucrats, loggers, students, retirees."[1] I can appreciate island ferries now—their slow, shuddering progress through narrow straights, past tiny islets covered in evergreens right down to the shore— but when I was a teenager all I remember is waiting. It often took a whole day to make the 90-minute crossing, most of

FIGURE 2 Ivy and Leo, Vancouver Island.

the time spent sitting in the car in an endless queue. Once on board the smell of the infamous burger platter, fast food sweating under a plastic dome, lingered in the corridors.

Islanders rarely see their island the way visitors do. I'm still surprised when I hear the things people get up to on the island where I grew up: whale watching or sea wolf spotting. Like most locals I went camping every summer, but I never went skiing or surfing, rarely hiked or ventured far up the island, and definitely never went whale watching. If you ask a Madeiran the best thing to do on their island, they will inevitably suggest hiking the ancient *levadas*. Do they ever do it themselves? Absolutely not! Swimming, yes. And going into the mountains for a barbecue. (The cultural theorist Stuart Hall recalls swimming and barbecues as the dominant memories of his native Jamaica: "swimming before breakfast, the water still as glass . . . or in the afternoon, riding the surging, spume-tipped—and scary—ocean waves at Boston Beach, followed by jerk-pork and festival barbecues."[2] They are the two principal pleasures I remember from Madeira as well.) The rest is for tourists. That doesn't mean hiking and whale watching are bad, it's just a different way of approaching the island.

Vancouver Island was a different place in the 1970s and 1980s than the exclusive enclave parts of it have become. I lived with my mother in public housing in a neighborhood with a lot of other single moms, as well as recent immigrants, recovering hippies, and draft dodgers. Loggers and environmentalists were frequently at war. Parades of Hell's Angels lined the island highway. Everyone looked newly arrived. The only people who were actually from the island, for more than one or two generations, were the indigenous people of what became part of Canada: the nations and tribes of the Coast Salish, the Nuu-chah-nulth (Nootka), the Kwakwaka'wakw (Kwakiutl), and many other distinct communities. They had mostly been pushed to the margins in the last century.

"It was gorgeous and claustrophobic," says a character of these islands in Emily St. John Mandel's novel *Station Eleven* (2014). "I loved it and I always wanted to escape."[3] When as teenagers we couldn't yet physically escape, we looked to music instead. J was always at the center of things; the night started at his place. Punk was thriving on the island, a weird insular scene that produced bands like Nomeansno and Dayglo Abortions as well as makeshift venues like the OAP Hall, which hosted Dead Kennedys, Black Flag, Meat Puppets, and Hüsker Dü in the 1980s, or the Rat's Nest, where Gary Brainless put on shows with local bands and hard-core legends like DRI and MDC in the airless, sweaty basement. Some bands had locally inflected names like Red Tide, the House of Commons, or Captain Cook and the Nootka Sound (a sixties garage band who wore tricorn hats). We took drugs and bounced around the confined space of the lower island like it was a padded cell, but we were luckier than we knew. Yes, Victoria was a silly little England, a provincial capital with mock-Tudor houses, rose gardens, lawn bowling and cricket, but looking to Britain meant that punk and post-punk arrived early and mutated with the island culture.[4]

An island can often seem like an untouched paradise, a tabula rasa. Captains Cook and Vancouver must have thought so when they arrived in their tall ships and saw majestic old growth forest and relatively few inhabitants who would soon be subjected to the cruel logic of colonialism. The slate was cleared with violence, so that these islands could be given new names. (A statue of James Cook was unceremoniously tossed into Victoria's harbor in 2021.) At a personal level an island may take on many meanings: it can feel like a place of happiness or grief, and those associations are largely beyond your control.

As always, I was overwhelmed by a range of conflicting emotions as the ferry pulled up to the dock. My mother was there to meet me: we hugged, and her body felt smaller than three years ago.

The next morning I was standing in a grove of Douglas fir and red cedar. On the drive over I kept glimpsing the sea sparkling in the distance around every bend, with a crown of Olympic mountains above the clouds. A light rain was falling as it always did, and everything was green and wet to the touch. About fifty people were gathered, most wearing the waterproof material that served as a second skin in the Pacific Northwest. I spotted Eric right away and hugged him. J's first and second wives were there, not far from each other, leaning into the shoulders of other bodies. It felt like a terrible high school reunion: all the people I used to go to gigs and sit in the halls with, together for the worst possible reason.

The rain fell harder, but no one noticed. As I stood there, jet lagged and swaying like a tall tree, I took in the damp, earthy smell and the sound of grief. One face after the next, all of them familiar but older, stepped up to hold a wet and crumpled page or wiped the screen of their phone to read the words they had written in the few days since J died, since he was murdered—though the word still feels strange to say or write—on this island. His mother, whom I remembered as a fiercely animated woman, was now a tiny wisp dressed in black, supported by a helper on each arm.

J was discovered unresponsive in the early hours of the morning. Eric called me to break the news. I remembered a text exchange with J a few days earlier, when he mentioned that he was going home to the island. His murder is still unsolved, and few details had been released. The neighbors heard popping sounds. Everyone was asleep. When they found him he had sustained life-threatening injuries, according to the press, and despite attempts by paramedics to revive him he died at the scene. It was being treated as a homicide, with no motive or suspects.

Someone handed me a beer. A Joy Division song was playing: not "Atmosphere," as you'd expect at a funeral, but "Disorder," which felt appropriate to the sense of sadness mixed with anger and confusion and even exhilaration we all felt being there together. I caught myself smiling because I hadn't seen so many friends in years: the paradox of funerals. We were never serious, so how could we be now? We joked and cried; we tipped and reeled like ships in a storm. Ian Curtis sang, "I've got the spirit / Don't lose the feeling" like a man possessed. All of us were thrown into disorder by a murder on an island where no one was ever murdered, or hardly ever. (According to Statistics Canada, 796 homicides were recorded for the whole country in 2021. In Victoria there were five. Eight months into 2021, J was only the second.)[5]

I stood at the microphone with some notes that I had written on the plane. I remembered eulogizing my father not far from there. I said that J would probably expect me to read some poetry. I mentioned his soft brown eyes and gentle touch, his effortless (truly effortless, as in no effort) thrift shop style, his paintings and sculptures and stage sets, as well as the art from old bones and junk that he made as a teenager. How he and his father, who was also an artist, looked like twins from different eras when they stood together. I recalled that it was eighteen years ago this weekend that I was married on Thetis Island, one of the happiest days of my life, and J stood beside me in his old suit, with his daughter on his shoulders, pretending her long hair was his hair (he was already going bald). Twenty years ago today we were watching the news of 9/11 unfold early in the morning at his house. I was visiting from Ireland and planned to take the ferry across to Seattle but the borders had closed, so we sat in his living room playing with the kids and watching the planes crash into the towers over and over on that other island. Twenty years to the day. Twenty-five years ago that month my father died and I was with him on the island. My father was the age we were now. He and J got along well. I told the faces in front of me that my dad's memory had never left me, I still thought of him every day. Not sadly, either: I remembered his humor and his stories, and I knew that J would live on like that in all of us. But saying it only made me feel his absence more—an absence that hit me like vertigo.

I told them about J and his new wife visiting Madeira, dancing and drawing on rolls of butcher paper in an abandoned house in the Old Town. With his first wife in Prague, talking on a payphone, one hand resting absentmindedly on the stroller where his daughter was splayed out sleeping. Friday nights at his house for bacon on a bun with J's dad cooking in a paint-splattered blue shirt. Their house so down to earth, unchanging, and yet a wholly and unapologetically artistic realm. Copies of the *Evergreen Review* from the late fifties and sixties stacked beside the toilet. J's patience and the way he spoke to my kids. I was always caught up in the details, I said, but he seemed elemental, like a rock or a tree. During the pandemic we met on Zoom for drawing sessions, where J dashed off iPad sketches. The last session was a week or two before his death.

His sister spoke next, her eyes dark with pain—the intimidating older sister with the cool friends and the record collection that was our whole musical library. Then his second wife, who was an old friend, devastated by this blow to their new life together; and finally his first wife, whose every word destroyed me until she stopped speaking and I took a long, slow, trembling breath. More people stood up to speak, more songs, more tears.

The next morning I went to meet my dad's side of the family. After browsing numbly in the aisles of Munro's Books on Government from force of habit, I headed to the beach to meet my aunt and my sister and my burly cousin with his shy smile and butterfly neck tattoo. I saw how few of us there were left, and how the beach with its scattered logs was the same as when I was a kid. Nothing ever changes on this island, I thought, but of course it wasn't true.

The unimaginable brutality of J's death hit me in flashes during the following days and in detailed, horrific dreams at night. I sat in ferry traffic waiting to leave the island as I had done so many times, and felt a wave of relief when the ferry rumbled and started to pull away from the dock and the smell of diesel filled the air.

In this frame of mind, alone again and heading home, I thought of D. H. Lawrence's poem "The Ship of Death," written not long before he died: "Oh build your ship of death, your little ark / and furnish it with food, with little cakes, and wine / for the dark flight down oblivion." J had no time to prepare, to build his ship of death. I imagined only the terror of a journey with no land in sight, as the chilling lines further on describe: "upon the sea of death, where still we sail / darkly, for we cannot steer, and have no port."[6]

A year after the funeral I received a package from the island. Inside there was a small envelope with "READ FIRST" written on it, as well as a larger one. I opened the smaller one and read that the larger envelope contained correspondence between me and J that was found in the house where he died, salvaged by his sister after a recent flood. Some of J's letters were replies he never sent. I couldn't bring myself to open the big envelope.

This message was the hardest to write. I'll end it by stuffing it into an old bottle of the kind of whiskey J liked to drink. It's made of clear glass, square with embossed lettering, and a gold tin cap that rattles when you spin it off.

3 ESCAPE

PORTO SANTO, SUMMER 2020

There were also good times on islands, happy times: before and after and between the bad times. After four months of lockdown, we needed a break. Shortly before leaving Madeira for good we took a final, anxious holiday to its sister island a ferry ride away. Madeira is an archipelago, like the Canary Islands, the Azores, and Cape Verde, all of which make up the region off the west coast of Africa known as Macaronesia. There are six uninhabited islands and sixteen islets in total, but only one other is inhabited: Porto Santo, a smaller, flatter, sandier island that was settled a year earlier than the main island, presumably because it was easier. Porto Santo's historical claim to fame, sold as fact by locals, is that Christopher Columbus married the governor's daughter and lived there for a year before setting out on more famous expeditions.

Madeira and Porto Santo are roughly forty miles apart, a passage that takes two and a half hours by the daily ferry that runs between them. The sandy island is the subordinate of the pair: poorer, drier, less powerful, and altogether less grand—but its beach is second to none. Nowadays it serves mostly as a holiday island for "mainlanders" (i.e., main islanders) who flock to it every August. The principal activities are swimming in the sea, lying on the beach that extends the length of the island, drinking beer at one of the

FIGURE 3 Kids diving off the pier, Porto Santo.

little palm-roofed huts, eating fresh seafood, and riding a bike, which is all but impossible on the steep slopes of Madeira.

The seas were very rough on that particular crossing, and everyone was feeling it. What happened next was probably the worst thing that could have happened during that first pandemic year, when everyone was sensitive to bodily fluids: all the passengers on the ship started throwing up. White plastic sick bags were scattered around the beige upholstered seating areas. Midway through the journey many of the bags were full. The bathrooms, if you dared to go in them, echoed with cries of misery and humiliation in several languages. Masks obviously did not help.

Ah, but once you're there!

Once you reach the other island everything is perfect, all pain is forgotten. Just sand and sun and sea: the kids off playing all day, lunches of fresh scabbard fish and steak sandwiches with mayonnaise on sweet potato flatbread, trays of fresh tuna and octopus in oil and vinegar and finely chopped shallots and herbs, grilled limpets with lemon and garlic, bowls of reddish-brown olives and gold lupins, and endless small, cold glasses of beer. The guy with a kiosk and a soft-serve machine who offers two confusing choices, year in and year out: lemon and chocolate or coconut and blackberry? Driving the dirty white Suzuki Sidekick along the straight coast road or on the winding blacktop through the inland desert, exploring volcanic caves and secluded rocky beaches, meeting friends for dinner with our feet dug into the sand, watching the kids dive endlessly off the pier. Nothing but this, day after day.

It was exactly what we—and everyone else—needed in the summer of 2020, temporarily freed from the extreme isolation of the first lockdown. In Porto Santo at that time there was almost no trace of the pandemic, far less even than on Madeira. There was no crowding, just open beaches and beach houses with lots of space between, everything happening out in the sun.

Although it looked empty, the island was almost as booked up as usual that summer and we had to settle for a rented cottage far from the beach. In the past we had stayed at the Hotel Porto Santo, which was fancy but affordable in the off season. The hotel looked the same as when it was built in 1970, right down to the small, cube-like televisions. The floors were a rich, exotic hardwood, and there was a pool and mini golf. Everything was very sedate, including the staff, who barely moved from the front desk. There was a lounge named after Manoel de Oliveira, the Portuguese film director who was as famous for his longevity and work ethic as anything, since he lived to be 106 years old and kept adding new films to his oeuvre until his recent demise. The director visited the hotel every summer. He had been a race car driver in his youth and won the International Estoril Circuit driving a Ford in the 1930s. The hotel was a classic and like the island itself was impervious to anything but the slowest pace of change.

Ironically, it was while staying at this hotel that I nearly suffered the quickest and most drastic change a person can go through: death. We were on a weekend break with friends. James and I set out for a swim with our two boys, already feeling sluggish from too much wine the night before. The sea was calm in comparison to the rough swells we were used to on the big island, where the fear of drowning was real and frequent, and before long we had been playing in the surf for more than an hour. Then James's son got out, leaving three of us. Without realizing it we gradually drifted down the beach, off the sandy bottom and onto a stretch of slate-like rock. Suddenly the waves were breaking faster and harder. Tourists drowned every year in those islands, just as tourists regularly fell off the narrow paths that ran beside the *levadas*, sometimes a dozen accidental deaths a year, but you hardly ever heard about it.

In an instant we found ourselves in trouble, and we all knew as soon as it happened. I was furthest out, then James a bit closer to shore, and then my son. James was calling to him. Then none of us said a word and there was only the sound of crashing waves and the taste of salt water. James tried to swim to my son, as did I, but all of us were immobilized as if in a terrible nightmare, trying to keep our heads above water. The thing that terrified me was seeing James, a powerful swimmer, in as much trouble as we were. Meanwhile on the beach nothing had changed. We struggled in silence, and no one noticed. We could see people lying on their towels, reading paperbacks, their children digging in the sand. What if James drowned and I survived, and I was held to blame for his death, just weeks after they had moved to the island? Or what if—more likely, I thought—I drowned and James saved my son, how bad would that look!

Finally, after what seemed like an eternal life-and-death struggle but probably was only ten minutes, the waves eased and we all made it to shore, crawling through the crashing surf. We walked back along the beach and found the others sprawled out on colorful towels with sunglasses and snacks. They hardly looked up as we approached. Flinging ourselves down on the sand, breathless and slightly melodramatic, we began to describe what had happened. My son said nothing. Our partners took our near deaths lightly, although I got a mildly disapproving look for endangering a child. We lay quietly for a few minutes. Then Clyde stood up, walked a little way off to the bushes at the top of the beach, bent at an awkward angle, and vomited. His mother called to ask him what was wrong. She turned to me and said, "What did you do to him?" James and I went to the beach bar and had a small, cold glass of beer, possibly the most delicious glass of beer anyone has ever had. The episode made an indelible impression on all of us, and colored not only subsequent visits to the island but also the next three years of my relationship with James, before he left Madeira and moved to Paris. My son eventually began to talk about it, and gradually it solidified into a narrative that he brings up whenever the conversation falls on death, drowning, or Porto Santo.

The idea often came to me (especially during the pandemic)—what would happen if I died on these islands? The protagonist of Thomas Bernhard's *Concrete* (1982) takes a trip to Mallorca after agonizing for most of the book about whether or not to go. He finds himself wandering through the cemetery in Palma, musing: "although I have always believed myself indifferent to the question of where I am buried, I thought to myself now: This is one place where I don't want to be buried."[1] So when we returned to Madeira I paid a visit to the British Cemetery. I had walked past its high stone walls every day on my route to work but I had never ventured inside, and it was on my list of places to see before we left the island. I was also feeling blue that day, and I thought that seeing all those dead people might cheer me up, since I myself was not yet dead.

In fact, it wasn't really a British cemetery, as the white marble plaque stated, but rather a foreigners' cemetery, probably meant for all non-Catholics. In earlier times, Protestants and other non-conformists who died on the island were simply tossed off a high cliff into the churning sea below. I noticed that, for a supposed British cemetery, the old British wine families were conspicuously absent, but I assumed they were buried by the private chapels on their own estates. I passed a grave that read: "Harry C. Stone, American. He believed in God." Not far from Harry, along a row of coffin-shaped cement graves, was the monument to a West African princess of the Yoruba people named Sara Forbes Bonetta. She had been captured and enslaved by colonizers but was later freed and apparently welcomed into British society and became the goddaughter and ward of Queen Victoria. She died, still in her thirties, during a visit to the island. Although her story was unique and captivating, it was how every story seemed to end: she came for a holiday and never left.

When I first entered the cemetery and saw the old broken gravestones and wandering vines bordering the paths, I felt a cold chill of terror at the thought of being buried there. To be planted in this random spot for all eternity, surrounded by strangers who were themselves strangers, all of us united only by our strangeness—what a terrible fate! As I walked on, however, I began to feel more at home in this quaint little tropical garden. One grave had a plaque that read: "These stones were brought over from Deserta Grande." The Desertas are neighboring islands in the archipelago, uninhabited except for some sheep and a rare species of wolf spider—by some accounts the largest wolf spider in the world. What an interesting thing to do, I thought, to go out in a boat and brave those giant spiders to bring back some nice stones for a grave.

By the time I left I could imagine no better end for an exile than coming to rest in this peaceful and cosmopolitan corner of the world. The thought crossed my mind: maybe there is still time to die here before I leave. And then, my mood having lifted as I'd hoped, I popped this message into a Brisa passion fruit soda bottle and walked home towards the *quinta*.

4 QUARANTINE

MADEIRA, SPRING 2020

Sometimes the inescapable feeling on an island is simply that you cannot leave. This was true during the Covid-19 pandemic, when the planes and ships stopped arriving and everything fell silent. The Italian writer Italo Calvino, born in Cuba, wrote that, "Islands have a silence you can hear."[1] During the pandemic it was deafening. At least we were safe on the island, I said, because the danger was across the sea. "Yes," Simone replied, "until it gets here: then we'll all be dead." She had a point. We became not only islanders but islands within islands within islands, inside the walls of the *quinta* and the house and our rooms, like photos I had seen of an island in a lake on an island in a lake on an island in a lake and so on. Our kids used to watch the Disney version of *Swiss Family Robinson* on DVD—a fantasy of mastery and survival in a strange and hostile environment. (There is a scene in which they hoist a quarantine flag to scare off pirates.) We had brought this fantasy to an actual remote island, but it was not until the pandemic hit that we felt truly isolated.

The *quinta* had sheep, chickens, old dogs who wandered freely looking for someone to greet, stray cats who fought with our cat, countless lizards, birds of paradise and frangipani, papaya and pomegranate and avocado and Surinam cherries. Our house was a whitewashed, red roofed former servants' quarters, with a private garden enclosed by an old rock wall and a view that overlooked the *quinta* and vineyards and the city and sea beyond. Because the whole island is terraced,

FIGURE 4 Bolacha Maria on the *quinta*.

the upper floor was the ground floor on the road above. On the *quinta* side there were three sets of French doors, left open year-round; we trusted the cat to keep the mice and lizards out. The owner was a bachelor in his forties with an English name and demeanor who inherited the property from his father, and who was the cousin of our friends and neighbors. The neighbors' house also sat along the top wall but was much grander than ours. The only other family we saw during lockdown were the Romanian intellectuals who lived in James's old house by the side gate; they spoke flawless English and had two shy, wild children. There were two large manor houses normally full of visitors which now lay vacant. We negotiated use of the swimming pool for the whole spring lockdown; it was bordered by a frog grotto and a faded tennis court. There was an ancient banyan tree that was large enough for the children of the *quinta* to play house. Everything was crumbling and overgrown in just the right way.

The fortified wine that Madeira is famous for, and which made a few English and Portuguese families fortunes they still possess, is distinctively an island wine. Islandness and remoteness were essential to its invention and unique process. The wine is distinguished by the particular varieties of grapes, terrain and growing conditions, but above all its character is defined by the long journeys it took overseas: to England, the East Indies, or the United States. The flavor of the wine comes from being transported across the ocean at high temperatures: it could not spoil, because spoiling was part of its character—it was already oxidized. Thomas Jefferson was famously a fan, hence the enduring claim that the signing of the Declaration of Independence and other symbolic moments in US history were marked with a toast of Madeira. The colonial period was its heyday. Like port and sherry it has fallen in and out of fashion over the decades, but living on the island I developed a special fondness for the wine still made by my friends and neighbors. All the major grape varieties (Bual, Malvasia, Verdelho, Sercial, as well as Terrantez) were grown in the *quinta*'s vineyards.

The pandemic diary I kept for more than two years begins on March 7, when an earthquake struck. This was a rare and unsettling experience, even though it was understood that such things sometimes happened in the region. (When the Great Lisbon earthquake of 1755 struck, decimating the Portuguese capital and killing tens of thousands, the Atlantic tsunami it caused hit Madeira first, followed by the Azores and the Canaries, and was felt as far away as the Scilly Islands off the South Coast of England, the West Indies, and Newfoundland.) The 5.3 magnitude earthquake happened at 9pm on a Saturday, cracking the plaster and causing us to run outside. My son and his friend were exploring a cave on the beach at the time, while the adults drank cocktails (the usual arrangement). The boys emerged to relieved cries. There was a tsunami warning, but nothing happened. A few days later the first confirmed cases appeared on the island, and unreality set in.

From one day to the next, everything stopped. All but a few planes stopped landing at Cristiano Ronaldo International Airport. Cruise ships stopped docking in the harbor. Our neighbors stopped inviting us to their parties. But the first lockdown was not entirely unpleasant. As the cartoonist Lynda Barry said later: "the pandemic introduced something that I had always fantasized about from the time I was little, which was being marooned."[2] Being marooned was a novelty, at least for now. Later there would be tragedy, friends becoming gravely ill—but not yet, for us, in this first phase. We felt ashamed of our luck. But luck was all we had: unlike our wealthy neighbors we owned nothing and saved nothing. We escaped from austerity-hit Lisbon to this remote island, on the most basic calculation, because it seemed affordable. What the institute couldn't pay us in salaries they made up in travel budgets and barbecues.

Ronaldo himself flew back and forth from Italy, where cases were rising sharply, to visit his mother who had suffered a stroke. The World Health Organization declared a pandemic on March 11, and on March 13 Donald Trump banned all travel to the US from Continental Europe. Tom Hanks and his wife announced that they had the virus in Australia. The NBA suspended the rest of its season. I talked to my mother on her far-off island, and we agreed she might have to cancel her visit in April. "It's fine, mostly," I said, "but the atmosphere is weird for a holiday."

Panic buying set in with the realization that the island had limited supplies. We found ourselves in a shouting match with a red-faced man at the checkout, who told us to go back to wherever we came from. Most people were calm and polite, but their carts were piled high. Cases surged in Spain as it became the second worst hit after Italy. Portugal banned travel to and from countries with high numbers. Porto Santo shut itself to residents only—no visitors from Madeira. On March 15 the Netherlands closed restaurants and schools, and Germany closed its borders. In a tearful conversation with my mother, she finally cancelled her visit. She told me about being hospitalized during the so-called Hong Kong Flu pandemic in London in 1968—a pandemic I'd barely heard of that killed a million people.

The next day schools finally closed. Pubs closed in the UK; Seville cancelled Easter processions; Rotterdam cancelled Eurovision; Ireland cancelled St. Patrick's Day celebrations. I went to the local shop up the hill and found the lights were turned off to save money, crates on the counter overflowed with local fruit, and the doorway was still crammed with old men holding bottles of Coral beer. I bought wine, canned beans, vegetables, flour, eggs. The idea of a very long quarantine, a year or more, began to circulate. Our house had movie nights, chore lists, a new home study schedule, an abundance of cooking and baking, and some heated arguments. Italy's death toll passed China's, and on March 20 California declared a stay-at-home rule.

When the first cases appeared on the island, the local government decided to quarantine arrivals for two weeks in a newly built *Truman Show*-style ersatz holiday village near the airport. There was talk in the papers about repatriating foreigners. I started carrying my residence card the rare times I left the *quinta*. I ventured out to a local clinic wearing a mask and latex gloves to get a prescription for an ear infection when it got so bad that I couldn't hear properly. On the way home I experienced the strongest reaction yet to my being an obvious foreigner, from a woman in the bakery where I stopped to buy a loaf of bread. She was visibly shaken, her back pressed against the wall—but another woman in the shop said reassuringly to me in Portuguese, "You speak Portuguese, right?" and I smiled, "Of course." The situation was defused.

By the end of March the US was leading the world in virus-related deaths, while Italy reached nearly a thousand per day. The British Prime Minister had the virus. Our kids followed school lessons in Portuguese on local television. We played badminton and did jumping jacks every afternoon under cascading flower blossoms, cut each other's hair in the garden, and watched all the *Rocky* movies. We grew increasingly bookish and insular, mostly happy in our forced solitude. I exchanged photos with James, now at his in-laws' house in Brittany, comparing baking and grilling techniques. One day we measured our cat: standing like a cat, standing like a human, hips, bust, nose to tail, wingspan, cranium, paw size, pants size. Emotions passed swiftly like clouds, sadness to irrepressible joy. At the end of March we agreed to meet the neighbors in the empty field below their house, drinking from our own bottles of wine and shouting the latest rumors.

On the first day of April I got a new job on the mainland. It was a strange time for good news; unbelievably good news in fact, a lifeline out of the blue. My wife had already found a job at the same institution, which meant we would leave in four months as long as everything didn't collapse in the meantime. It looked very possible that everything would collapse. EasyJet grounded all flights in Europe. The death toll in New York started to spiral out of control. There were a million infections worldwide, then two million. Cases in Madeira remained low, around fifty, with only a handful on Porto Santo. There was an outbreak in a housing project in Câmara de Lobos that started with an Easter celebration, and as a result a "sanitary fence" ringed the village for weeks. The islanders remained surprisingly pro-lockdown through it all. Pain, self-sacrifice, and being at the mercy of larger forces were familiar to them. It was a chance to be truly insular, to cut ties with the mainland and live the dream of self-sufficiency. It was an island of plenty, with enough fish and fruit and wine for all.

Soon there were ten million cases worldwide. I remember standing in the garden, watching swallows and kestrels swooping overhead, lizards chasing each other and the cat chasing lizards, monarch butterflies drifting on the soft June breeze, songbirds singing long intricate melodies to each other across the canopy of trees and bushes, flowers of all colors and sizes spilling over stone walls, the scent of lavender and rosemary and jasmine and honeysuckle. A pack of wild dogs got in one night and slaughtered most of the sheep and chickens. The landlord bought more sheep and put up a higher fence. "Be sure to keep all the gates to the outside locked," he warned us.

There were riots in Minneapolis after the killing of George Floyd. The President's tweets were taken down for inciting violence. Protests erupted across the Americas and Europe and around the world. Meanwhile it was calm on the island, with almost no new cases. Restrictions began to lift as summer approached. I received an official email telling me that the State of Emergency had been replaced by a State of Calamity. I awoke to the sound of a loudspeaker: a man was driving slowly through our neighborhood selling fresh seafood out of his van. I considered flagging him down to buy some limpets. Did we really want to leave?

Before we embarked on a last holiday to Porto Santo that summer, we had our first Zoom drawing session with friends in Canada. It was originally intended as something for the kids to do, but it quickly became a support group for adults. Since I was on European time I often had a beer in my hand; the others drank coffee. The drawings were everything from pencil and marker on scrap paper to sophisticated iPad drawings. The sessions continued every week or two over the next year, right through the summer of 2021.

As we packed up for our holiday and impending move, with everything in boxes and the cat looking nervous, I thought of what I would miss: the sounds from the school on the hill, the smell of diesel, the air and the sunlight in the garden, the colors of orange blossoms and red tiled roofs, the landlord standing by the gate, his firm handshake and distant yet confident eyes (the gaze of eight or nine generations cultivating this very piece of land), the stormy north coast and the laurel forests, the peaks and surreal plateau in the clouds, the strange institute where we worked, and our own Doctor Moreau, standing in our garden after a long dinner, smoking a Cuban cigar and drinking a last glass of whisky before driving off into the night. Rolling up this message I slid it quietly into a bottle of Blandy's vintage Terrantez that I found in the recycling bin by the front gate.

5 HOLIDAY

BELLE ÎLE, SUMMER 2019

As I write this next message in a bottle—an old pastis bottle, as it happens—I'm listening to "The Pork Sausage (An, Anduilhenn)," a scratchy, jaunty tune from the 1957 Smithsonian Folkways album *Songs and Dances of Brittany*. The inspiration was a short film my son made about Houat, an island three miles long and less than a mile wide, which he visited last summer. He interviewed locals about the island's history, the trip to and from the mainland across the Gulf of Morbihan, and changes they had witnessed in their lifetime. Before the pandemic that led to our move, we were sometimes lucky enough to go on island-to-island holidays. One blissfully unremarkable journey led us to Belle Île, in the same area as Houat. We met up with another family on the mainland—not just any family, but our former neighbors on the *quinta* in Madeira, who lived in the house the Romanians later took over, and with whom we shared countless nights by the fire with bottles of wine. We met James and Elise and rode the ferry from the crowded port of Quiberon, with its faded seaside charm, to the misty forests and rocky beaches of the lush little island.

Even more than Porto Santo, each day here was exactly like the last, and the next. The pleasures were cheap and plentiful. Every morning I walked forty minutes from our cabins down to the village to buy fresh bread, croissants, and *pain au chocolat*. On most days I had to wait for a bridge that was raised to let a sailboat pass through, and I stood there feeling unhurried and unbothered, smiling at the people on the boat, who smiled back. After the boulangerie I would stop for a *café crème* at the *bar-tabac* next door, half a dozen baguettes *de tradition* standing up between my feet, reading the rumpled newspaper on the counter and trying to translate the gossip I heard around me. I was not in a hurry there or anywhere I went. Everyone was on island time.

One morning I was listening to "Evidently Chickentown," a discordant rant against England by the Mancunian punk poet John Cooper Clarke ("The bloody pies are bloody old / The bloody chips are bloody cold"). I was thinking of the stark contrast between islands in the sun (like mine, then) and islands of bleak unceasing rain and impenetrable gloom (like his, where I'd spent years of my life). Belle Île was a bit of both, weather-wise, but it was evidently *not* chicken town. It rained heavily and fairly often, but when it did you could simply stand under an awning until it passed, in summer at least; when the sun came out, you could sit on an ornate old bench till your clothes dried. The food was exceptional, the drinks were good. Sometimes the morning bread run took me hours, and the others would ask why. But what did we have to do? There was always coffee and bread and butter from the previous day to tide them over until I returned. Later we would walk to the beach, walk home again, open a bottle of wine, let the kids run around, send the kids to bed, open another bottle, return to our cabins, make love, fall asleep. The same thing, day after day, and we all wished it would never end.

My suitcase was half full of books, paperback novels by the prolific Belgian crime writer Georges Simenon. They were short and simple and atmospheric, perfect holiday escapism. The characters in his novels spent most of their time ordering aperitifs in dingy cafes. I started with *The Mahé Circle* (1946) because it was set on an island, although Porquerolles is in the South, just off the Côte d'Azur between Marseille and Cannes. There is a strong sense of place: the heat of the sun, the smell of Mediterranean cooking, the sound of birds and cicadas. Unlike his more popular books featuring the detective Maigret, this is one of the dark, philosophical *romans durs*, written during the Nazi occupation. The reader's pleasure comes from a feeling of cozy safety and security against the bleak narrative. The pages of the book were soon full of sand.

The provincial doctor Mahé somehow manages to make even watching a game of *boules* into an existential crisis.

He succumbs frequently to a Sartrean nausea, staring into the murky depths of the sea or at the five o'clock shadow on his friend's cheek and his mouth as it speaks meaningless syllables. He has a terrible time on his first visit to the island, yet he decides to go back the next summer. He needs to escape the malaise of his settled mainland life, into the wild unknown. Fernando Pessoa wrote: "I feel like fleeing. Like fleeing from what I know, fleeing from what's mine, fleeing from what I love. I want to depart, not for impossible Indias or for the great islands south of everything, but for any place at all . . . that isn't this place."[1] For Mahé it is similar, but it is clearly a mainland-island dialectic. On the mainland, "Everything was in its proper place." On the island, in contrast, everything is hostile and chaotic, "a kind of life that was too vivid."[2] He becomes obsessed with a young woman in a red dress, the daughter of a soldier who leaves the island to get drunk to avoid the prying eyes of his fellow islanders. She represents mystery and escape—from the Mahé Circle, the trap of a bounded, predetermined life, represented by the mainland.

He keeps returning every year, and by the third time he has been somewhat accepted: he fishes with the locals and plays boules with the old men. (Although Mahé gradually acclimates to island life, his wife does not: she takes the children daily to the beach, but never thinks of putting on a bathing suit, rarely looks up from her sewing, and sits with her back to the sea.) Simenon depicts something essential about the island ideal, especially the South as seen from the

North, as a place of exotic otherness. He visits in summer, when the island casts a magic spell; in winter it enters a sort of quiet hibernation. It is a dream, but one that is always accompanied by feelings of strangeness and unresolvable ambivalence. The island is both the remedy for existential malaise and its embodiment: "He both longed for [the island] and dreaded it. He knew that he would be unhappy there, that he would feel an outsider," yet he goes anyway.[3] Islands have always served as a blank screen for projections of desire, fear, and other powerful emotions; or, to shift the metaphor slightly, an empty stage to act out all sorts of interpersonal conflicts and internal struggles. In the end, however, he resolves nothing of his affairs, and escapes instead into the sea, which swallows him silently.

In contrast to the book, which captured something of my ambivalent relationship with Madeira, my time in Belle Île was uncomplicated. There were the usual minor dramas, but overall the memories are pleasurable: light, hazy and scattered, like clouds in summer. The memory of James walking down a narrow path with his youngest daughter on his shoulders. Ripe green figs with red insides lined up on the porch outside the cabin. The bakery with its ridiculously buttery *kouign-amann*, a Breton pastry that will kill anyone with a weak heart. The daily routine of making baguette sandwiches for nine people, packed in cloth bags along with madeleines, bruised apples (never eaten but taken along for form's sake), bottles of water and lemonade, a flask of coffee, butter biscuits. Houses with shutters painted blue, the same blue of the rowboats in the harbor and the coveralls worn by the workers in Port de Le Palais where the ferry docks. (In the novel Mahé thinks: "he'd buy some blue canvas trousers . . . the bright blue that made such a sumptuous patch of color in the sunlight."[4] Levi's might be quintessentially American, but denim, *serge de Nîmes*, has its origins in France.) The perfect, cold blue sea. German bunkers still scarring the beaches. Oysters and crêpes in the market. Bowls of hot chocolate for the children in the cafe after they'd turned bluish from staying in the sea too long. All of this, until it was time to return to so-called reality on another unreal island.

The English writer Esther Freud once made a pilgrimage to Tove Jansson's tiny island in the Gulf of Finland, the setting of *The Summer Book*. At first she found it claustrophobic— it took exactly four and a half minutes to walk around the island. But before long her sense of time and space shifted dramatically: "I have finally arrived at island time. . . . My focus has changed now. The island is no longer quite so small. The rocks have become cliffs, the creek a ravine."[5] In the book itself Jansson describes how the scale gradually shifts, so that every last twig was not only in its proper place but given a place of importance relative to the scheme of things.

As a designer, James has taught me the importance of constraints: of time, materials, cost, function, usability, size, integration and compliance with existing systems, and so on. Constraints are the essential starting point for any project. Far from hampering a good designer, he said, constraints should act as a spur to creativity. Islands, especially small islands, focus the attention and constrain the scope of action. This can be a source of frustration if you are a teenager trying to explore the wider world, as I was on my island in Canada—an adolescence spent waiting for the ferry to freedom. But in a chaotic world, islands can be a refuge precisely for their limited nature—*just this, just us*—this bakery, this beach, these friends, this sea.

6 LONGING

SÃO VICENTE, SPRING 2019

I went to meet my son, now a streetwise and slightly scruffy college student, in Rotterdam. While it is not usually thought of an island, much of the city lies substantially below sea level, behind dams like the one the city is named for (the Rotte dam), and parts of it are islands in a river delta connected by a network of tunnels and bridges. He suggested Tia's as a meeting place because it was equidistant from downtown and Delfshaven, where he lives. In the Dutch Golden Age cities like Amsterdam, Haarlem, and Delft grew rich off transatlantic trade, including the slave trade. Delfshaven, as the name hints, was actually the harbor of inland Delft at one time, and the departure point for slave ships (as well as the Speedwell, the ship that carried the Puritans from Leiden to England, where they transferred to the Mayflower).

The cafe was run by a woman from Cape Verde, the eponymous and formidable Tia (auntie), and it served good Cape Verdean and Portuguese dishes as well as bottles of Super Bock beer. When I got to the place my son was already there, his long frame draped lankily over a dark wooden chair, chatting politely with the owner in Portuguese. They turned and she welcomed me in Portuguese, then asked my

FIGURE 5 Mindelo, São Vicente, Cape Verde.

son if I understood. "*Mais ou menos*," he replied with a smile. "He's better in English."

We ordered a cold beer in a tall thin glass and a couple of *pasteis de bacalhau* as an appetizer. When the main dishes arrived, mine was a hot plate of *cachupa*, a signature dish of Cape Verde, in this case the delicious fried version more often eaten for breakfast than lunch. I could understand why it made a hearty breakfast—it looked like a large plate of grains surrounded by chunks of meat and topped with a fried egg. It went well with the beer and a few lashings of hot sauce. My son continued his conversation with the owner intermittently while we ate, replying to her in Portuguese over his shoulder when she asked if we were enjoying everything and if we'd like another beer (yes please). He told me the place was fairly sedate except when there was a football match featuring big Portuguese teams like Benfica and Sporting, which also had dedicated followers in Cape Verde and Madeira.

I remembered a work trip from Madeira to the island of São Vicente, shortly before the pandemic, accompanied by two male Portuguese colleagues with whom I had little in common. It was related to a shared research project between our institute and the university there, but the other two were involved in admin rather than research and showed no interest in the presentations we attended over the next few days. Their job was to secure more regional and European funding, and they spent their off hours going out drinking in the island's main city of Mindelo where we were staying. I was only vaguely connected to the project myself, since it was about sustainable energy and I was trained in modernist literature. But I intended to learn as much as I could about the island, its energy, and its people. Before arriving I only knew that the famous *morna* singer Cesária Évora came from Mindelo, and that she had recently died there.

The first thing you notice when you step off the plane onto the dusty tarmac at Cesária Évora International Airport is dazzling sunlight, followed immediately by the palpable absence of water. You sense it as if by animal instinct; it dominates your thoughts as you race across the barren moonscape on a two-lane road in an old Mercedes taxi, chasing an endless series of shimmering mirages. A few tiny pink flowers struggle at the roadside, but otherwise everything is dust. It's hot and dry, in short, and the sight of the deep blue sea on every horizon only makes you thirstier. The lack of water haunts every aspect of life on the island, and you feel it with every privileged sip you take of bottled water or Coke. Although Cape Verde shares the region with three European archipelagos—the Canaries, the Azores, and Madeira—this is clearly Africa, and there is no financial buffer sent from mainland Europe to cushion the reality of life here.

At the same time, life is in many ways good, and better than other parts of Africa. It is a democratic country, crime rates are low, education rates are high, and tourism is booming. Women pass by carrying enormous bundles of produce from the market on their heads, completely unselfconscious and elegant in their stride. In the afternoon to escape the heat I walked along the city beach, my feet immersed in turquoise water, and a small pack of dogs ran alongside me, excited but unthreatening. They stopped to chat with another dog, stretched out on the sand by herself like the human sunbathers. I stopped by a floating bar at the pier, an international zone of sailors and tourists of all ages and nationalities drinking bottles of Strela under sail fabric, and took in the easy constant creaking as the platform rose and fell in the sheltered bay, the scent of the sea and fried fish, the boat maintenance men in sunglasses and flip flops moving to and fro. Eventually I returned to the hotel and took a long nap.

I found the afternoons unbearably hot, but the evenings were perfect: from dusk onwards everything became muted and sultry, with soft winds and a relaxed, pleasurable pace. At around nine o'clock I met my colleagues for dinner at an outdoor restaurant that served fresh tuna and a flinty red wine called Sodade from the neighboring island of Fogo. From there we went to a bar with live music, where I drank several glasses of a concoction made from rum and molasses that was as thick and rich as it sounds (not to mention efficient, since both ingredients are derived from the same cane sugar). We sat with the band afterwards and I spent most of the conversation smiling contentedly, following only the general flow, while my colleagues interrogated the musicians about the history of *morna* and their unusual instruments, and how Portugal (which must always be reflected in a powerful and favorable light) might have played a role through *fado*, just as they claimed to have given Hawaii the ukulele. My colleagues talked about the alarming rise of anti-depressant use in Madeira, and how people in Cape Verde looked much happier although they were materially less wealthy. The young black musicians shrugged and smiled.

The next morning, slightly hungover from the rum and molasses, we met in the lobby to be picked up by a representative of the university in what the owner proudly proclaimed to be the island's first electric car, a dusty white BMW. I wondered aloud where he charged it, but his Portuguese was too fast for me to follow, and I turned instead to look out the window while my colleagues engaged him in animated conversation. At one point we passed a wind farm where the large turbines had only one or two rather than the usual three blades, and everyone laughed—an energy joke. Soon we arrived at the campus on the edge of town, which looked parched and deserted. I gave my presentation on the ideas James and I came up with for renewable energy in the region, which was met by polite applause but mostly ignored by the engineers who preferred realistic plans over sketches of imaginary infrastructure.

Like Madeira, the Cape Verdean volcanic archipelago was uninhabited until it was claimed by Portuguese colonizers in the fifteenth century, and it was used as a convenient stopover point in the Atlantic slave trade. Unlike Madeira, it was populated by a majority of Africans rather than Europeans.

Charles Darwin made the islands his first stop on the voyage of the Beagle in 1832, noting in his diary the hazy dust-filled air, the laughter of the people, and, arriving directly from his own rainy and verdant island, "the novel prospect of an utterly sterile land."[1] As Antonio says to Gonzalo in *The Tempest*, the island has everything—"save means to live."[2]

In the market I visited alone on Saturday morning I noticed that one image followed me wherever I went. Every t-shirt, bag and beach towel was printed with the smiling face of Amílcar Cabral, the Che Guevara of Cape Verde, the poet and liberation leader who was assassinated in 1973. In his poem "A Ilha" ("The Island," 1946) he wrote about the pain of his "forgotten" island, like an abandoned and neglected mother:

> *Tu vives—mãe adormecida—*
> *nua e esquecida*
> *seca,*
> *batida pelos ventos,*
> *ao som de músicas sem música*
> *das águas que nos prendem . . .*

> (You live—sleeping mother—
> naked and forgotten,
> barren,
> battered by the winds,
> to the sound of songs without music,
> from the waters that confine us . . .)[3]

Cabral led a long and successful insurgency in Cape Verde and Guinea-Bissau, another former Portuguese (and French) colony on the nearby mainland, and was killed on the eve of independence as the colonialist dictatorship of Portugal's Estado Novo crumbled after the death of Salazar. Although Cape Verde has done relatively well since independence, emigration is still a defining characteristic of the archipelago, as it has been historically in Ireland, Scotland, Madeira, and the Azores. It can be seen in places like Tia's in Rotterdam, where expats gather for a taste of home.

One of Cesária Évora's most popular songs is the haunting and blues-like "Sodade," written in the 1950s by Cape Verdean composer Armando Zeferino Soares about emigrating to São Tomé and Príncipe, another (now former) Portuguese colony and island nation, and missing his home island of São Nicolau: "Se vou escrever muito a escrever / Se vou esquecer muito a esquecer / Até dia que vou voltar" ("If I'm going to write, much to write / If I'm going to forget, much to forget / Until the day I return"). The song is named for the Cape Verdean creole version of the Portuguese *saudade*—the famously untranslatable word for nostalgic longing for a place (usually the place one is from and had to leave) or a person (especially one who has left, like those who emigrated or were lost at sea). *Sodade* or *saudade* is a frequent theme, even the dominant mode, of both *morna* and *fado*. In fact the feeling those terms and musical styles evoke, although it has a special relationship to islands and coastal areas with a history of emigration or exile, is described in Shoshana

Zuboff's book *The Age of Surveillance Capitalism* (2018) as a feeling that is now universal: the dislocation from, and longing and nostalgia for, a feeling of home in the digital age.[4] I think of the lotus-eaters in Homer's *Odyssey*: "Any crewmen who ate the lotus, the honey-sweet fruit, / lost all desire to send a message back, much less return, their only wish to linger there with the Lotus-eaters."[5] The term "lotus-eater" is often used to describe the idle rich, and Charmian Clift uses it to describe the aimless bohemians of Hydra; but it is also a defense against *saudades*, a form of self-preservation. For what could be worse, but also what could be better, than to forget one's home and loved ones? It is a merciful fate for those who choose to leave—to never be plagued with guilt or longing. This message goes into a bottle of *grogue*, the sugar cane aguardente that was nearly my downfall.

7 EXPERIMENT

EDAY, FALL 2017

All islands have their own ecology, some more radically divergent than others from the mainland. At the same time islands can also be seen as smaller-scale versions of larger ecologies, and they can help us to understand larger systems—as I discovered working on a sustainable energy project in Orkney. The archipelago situated off the northern tip of Scotland is made up of twenty inhabited islands and around fifty more uninhabited. We were set to visit one of the smallest inhabited islands, Eday. Our Scottish connection Laura suggested it might be a good place to do some experiments. This letter, tucked into an empty Scapa whisky bottle, describes a process not unlike an episode of *Scrapheap Challenge*, with the idea that islands might teach us something about harmonious living.

James often quotes a passage from John Steinbeck's *The Log from the Sea of Cortez* (1951) that includes the line: "It is advisable to look from the tide pool to the stars and then back to the tide pool again."[1] Similarly, in Virginia Woolf's *To the Lighthouse* (1927), one of the Ramsay children plays with scale and godlike powers in a rock pool: "she changed the pool into the sea, and made the minnows into sharks and whales, and cast vast clouds over this tiny world by holding her hand against the sun."[2] In their boundedness, isolation, and scale, islands make perfect testing grounds—from Bikini Atoll to *The Island of Doctor Moreau*. Islands are often treated as living labs, from agriculture ("seed islands") to housing and policing. The term and concept of the "living lab" is still popular in academic and other contexts—but does anyone

FIGURE 6 James building a gravity battery with the community on Eday.

really want to live in a living lab? Islands with colonial histories are rightly suspicious of those who would use them as testing grounds. On the other hand, what if the "lab" is run by the island's own inhabitants, for the good? As I found on Eday, where everyone is an expert in renewable energy, this can sometimes be the case.

James, Enrique, and I left Madeira early on a Friday in October, bidding a temporary farewell to sunshine. We had won a sustainable energy prize with funding to conduct an experiment on another remote island. Enrique was originally from Lanzarote but had come to Madeira to study with James. The fourth in our party, Mohammed, a Scottish designer then living in rural Sweden, met us that night in Inverness. By Saturday morning we had reached the town of Kirkwall, on the Orkney Mainland (mainland here denoting the largest island in the archipelago). Laura arrived on the next flight, and filled us in on the place and the people since she knew Orkney well. Over a lunch of fish and chips we shared our thoughts about the gravity battery, the thing we were supposedly there to create, including what sort of scrap we might use to make it. There was also the question of what we should do with the energy it released. Previously we had powered a record player; this time we had in mind a lamp, or an old radio playing the BBC shipping forecast. None of us had ever been to Eday, which is an island of ten square miles with a population of just 130 people.

The afternoon ferry from Kirkwall to Eday was cancelled, so we headed over to Stromness on the far side of the

Mainland to spend the night. Enrique commented on the irony of Silicon Valley's dreams of colonizing Mars when we were stranded ten miles from our destination by a bit of wind. But we made the best of it, and after a full Scottish breakfast with haggis on Sunday morning we drove back to Kirkwall and caught the next ferry. We arrived on Eday after dark, windswept and salty, and followed the island's only road to the only accommodation, the hostel. The trip north had taken three days, leaving only three days to build before the planned demonstration.

We woke Monday morning to the sound of a large wind turbine spinning fast outside the hostel, telling us the weather conditions. In fact the island grid is powered entirely by renewable energy; Eday's experimental and community-driven use of renewables, including wind, tidal, and solar, as well as storage in hydrogen fuel cells, is the main reason we were so keen to visit. Even the electric heaters in the hostel were powered at certain times by energy overflow from the wind turbine outside. Everyone we met on Eday was extremely well versed in energy generation and storage, including the children of the local primary school who spoke knowledgeably about electrolyzers and curtailment.

We met Clive, our local fixer and project partner, after a breakfast of porridge and bacon butties. Clive was legally blind and in his mid-seventies but full of gusto and an extremely capable fellow. He spoke with a Cockney accent and hinted with a mischievous smile at his past adventures. He had seen every famous band in England play from 1965 to the mid-1980s and worked as a roadie for a while. Before we set out to gather scrap materials and tools, he gave us a pep talk: "This may look like chaos, lads. But I assure you the machine will be built, it will be demonstrated, and you will leave happy on the ferry Friday morning." Talking to Clive, I was reminded of Byron's satirical epic *Don Juan* (1824):

> . . . Man
> In islands is, it seems, downright and thorough,
> More than on continents—as if the sea
> . . . made even the tongue more free.[3]

In the car he pointed out the island's landmarks—his lack of sight did not hinder him on such well-bounded terrain—and mentioned some of the people we would likely meet that day. As far as we could tell you were not allowed to have the same name as anyone else on the island: when a second Kate arrived she was renamed Katie, and the second Mike became Mick. (This is as good a definition of a small island as any.) Clive warned us that the community would need some convincing before they got involved. We should expect questions like, "What's in it for Eday?"

The first stop of the morning was the Old Church, which had been bought by a woman from London in the 1980s and had sat derelict ever since. Here we found an old motorcycle, a red Kawasaki, parked in the middle of the church among other scrap (Super 8 camera, record player, typewriter). The motorcycle had only 12,000 miles on the odometer, but it was buried under a thick blanket of corrosive pigeon shit, and its insides were seized beyond reasonable use. At the second stop, the New Church one minute down the road, we found a large brass bell salvaged from a sunken steamship. We thought we might use it as a weight for the gravity battery. Permission would have to be sought, Clive said. The third stop was a yard full of rusted cars and engine parts. Someone asked: "Did he die?" Clive replied: "No, just left the island." James opened the hood of an old BMW and a rabbit jumped out.

We looked at the building site next, which was a shed by the pier used mostly for deliveries. It had a forklift and plenty of room, making it the first real success of the day. We would have to clear out when the ferry docked, but aside from that it was ours. We agreed it would do nicely. Our other Eday contact, Andy, who worked with Clive, met us at the shed. From there we drove to an old quarry on the far side of the island, a three-minute journey. As we walked around the site, staring up at the sheer sides from the quarry base, someone had an epiphany about making a gravity-powered keyboard. Andy, it turned out, not only knew how to play (he was currently the church organist), but he had also been the keyboardist in a 1980s band called Freeez, with three "e"s. They had a number one single in the US dance charts with "IOU," which also featured in the breakdance film *Beat Street* (1984). We all agreed that if we could get hold of a scrap keyboard the issue of what to do with the energy released by the gravity battery was solved.

After lunch we met some people from the community in a building next to the island's only shop. The key moment in this meeting was the suggestion that Mick, who was spotted walking past outside, had an old motorcycle in his barn; someone ran out to talk to Mick and he kindly agreed to let us follow him home. He was a large man in an old CCCP t-shirt, who told us in a Liverpool accent to mind the ducks and sheep. He opened the barn and dragged out an old dirt bike, its wheels clogged with hay; he used an axe to free up the front wheel, and four of us rolled it up the driveway in the rain and waited as someone found a van to bring it back to the pier shed. We were cold and wet, and the light was fading on our first day, but we had a motorcycle and a rough idea of what we were going to do on Eday. We ate a hearty dinner at Roadside, the island's former pub turned occasional restaurant (actually just a cozy dining room in a private house), and then returned to the hostel to drink whisky and go to sleep.

On Tuesday morning we arrived at the shed at 9:05 to find several islanders already waiting in coveralls, ready to work. We introduced ourselves, made some tea, and set up to start cutting into the bike, while Clive got on the phone to order a new chain from the Mainland. We quickly sourced some necessary tools from generous community members, including an angle grinder, a lathe, and a socket set, and got to work. (Part of the challenge was bringing nothing to the island.) By lunch the bike was stripped, leaving only the parts necessary for the gravity battery—the frame, engine, and rear axle. In the afternoon two of us went to the school to hold a workshop with the children while the others stayed back at the shed. The wind blew and the rain poured down. Countless cups of tea were consumed. Soon the day was over, the children with their thick Scottish accents went home for dinner, the shed was locked up, and at the hostel Mo made his mother's dhal. It was Halloween night on a remote Scottish island, so of course we watched *The Wicker Man* (1973). More whisky was consumed and we went to bed late.

The challenge on Wednesday morning was how to get the gravity battery over the fence and down into the abandoned quarry, our chosen site for Thursday's demo. We noticed a large tractor: who did it belong to? Could somebody drive it there? Health and safety were still a headache that Andy was dealing with, negotiating with the property owner in England and the insurance company. It was blowing a gale all the previous night and all morning; the rain beat down on the corrugated iron roof of the shed, making it hard to hear anyone speak.

On the positive side, people from the community were starting to get excited about working together on this strange project. Old habits were becoming unsettled as people from different parts of the island who rarely spoke to each other met and pitched in as a team. Hamish and Mel, both native Orcadians, came to join in and brought their teenage son, an apprentice engineer. An old Casio keyboard was found, and after some minor tinkering Enrique brought it back to life. The chain arrived in a padded envelope by the afternoon ferry. Calculations were underway for rigging a pulley over the quarry edge. More people showed up, to work or to watch. Clive told us stories of London in the 1960s, working in Carnaby Street and seeing the Stones in Hyde Park. "What brought you up to Eday?" I asked. "Cheap, innit," he said with a wink.

On Wednesday afternoon we went to use the lathe in the shed of a man named Mike, another Englishman and ex-submariner who lived in the old schoolhouse. Mike left a note in the shed telling us what to do if a blackbird showed up at the door—he had trained the bird to come in and ask for food when it was hungry. Sure enough the bird showed up, looking at us expectantly until we passed it some raisins and a biscuit. When we finished our machining Mike invited us into the main house. In what turned out to be one of the week's highlights, Mike led us through a hole in the wall to reveal no less than a full-sized model of the inside of a submarine, complete with salvaged periscope, control panels, and torpedo launchers. We walked through room after room, through sleeping quarters with life-sized mannequin sailors, until we reached the end and emerged back into the schoolhouse. We shook hands with Mike, amused and somewhat bewildered by what we'd just seen, and returned to the pier shed. An island is a refuge for eccentrics: it is where you go to do weird things you can't do on the mainland, like recreate an entire submarine in your house, or burn a million pounds for art, as Bill Drummond and Jimmy Cauty of British electronic band the KLF did on the Scottish island of Jura in 1994.[4]

That night at dinner we envisioned our trip to Eday as a three-act play. The first two acts, we decided, had established the principal characters and their relationship to the world they lived in. The inciting incident of Act One was our arrival on the island, with a mad plan to build a gravity battery using only scrap. The rising action of Act Two was the first three days of building, where we pitched in with the community to make the thing we'd set out to make—the prospect of public humiliation driving us forward. The character arcs of ourselves and everyone in the community developed under this pressure; Andy and Clive even told us that relationships between community members had been altered for the good by our presence. Old feuds had been put to rest. For our part we gained insights about ourselves, the nature of our work, and islands in general.

Every story needs a climax, and it usually involves overcoming a crisis. So it was not unexpected that we should receive a phone call at dinner that night, the night before the public demo, telling us that the absentee landowner would not allow access to our chosen site, the quarry, without insurance—and negotiations with the insurance company had reached an impasse. Andy was trying his best to provide evidence of due diligence to both parties, but insurance is about predictability and is naturally risk-averse. Testing a gravity battery made from scrap in an abandoned quarry on a remote island in adverse weather conditions with schoolchildren present is not an ideal scenario from the insurer's perspective. How could we bridge the gap between health and safety, on one hand, and daring innovation and experimentation, on the other? How would we achieve a satisfying result and leave happy, as Clive promised, on Friday morning?

Andy, as I mentioned, was a professional musician in his previous life—he had played on *Top of the Pops*. So when he sat down to rehearse for the demo on Thursday morning, now in his fifties, perched on a wooden crate in cargo shorts and hiking boots and a down jacket, tapping at the keys of a salvaged Casio, the other islanders laughed and jibed good-naturedly. One man shouted: "You've come a long way, Andy!" By 10 a.m. the crisis at the quarry had been bypassed. Just as the insurance issue was looking close to resolution, we decided it would be easier anyway to stage the demo at the pier shed, so that no major moving of equipment would be necessary. Now momentum was gathering towards the final event.

The school bus pulled into the parking lot in front of the pier shed and the children got out. They lined up and presented drawings of the gravity batteries they had designed earlier in the week. One child broke down under the pressure and sobbed loudly, but eventually held up his drawing between shaking hands. The weather cleared up. Everyone stood facing us in a semicircle and waited for the show to begin. At the last moment James improvised an attachment to an electric drill and used it to drive a super low-gear winch, slowly raising the water container that was the falling mass.

It seemed like a bit of a cheat—though in fact it wasn't, since the island's grid was powered by renewable energy—but the mass was now suspended, the energy stored until needed. We called for everyone's attention, released the mass, and the keyboard came to life. Andy played "Ode to Joy," followed by the theme tune to *The Flintstones*. The performance lasted several minutes, then the keyboard fell silent as the slowly descending water container touched the ground. The crowd went wild. We did it again, and then again—the last time letting the kids take turns banging out some noise on the Casio. It was a success. We cleaned up as dusk fell.

That night there was a party at the community center. Andy led a discussion with us and some of the islanders, including Hamish and Mel and their children, Submarine Mike, Ivan and his son, and a few others. It verged on emotional at times, as people discussed past achievements like the installation of a massive community-owned wind turbine, and the climate changed future of the island itself. We tried to impress that, given ideal conditions, Eday's energy storage solution would not be a gravity battery—which was something we had created out of the particular materials and terrain of Madeira—but rather a bespoke solution for their island, built not from Madeiran *sucata*, or scrap, but from Orcadian *bruck*, like a flow battery made with seawater. But we had three days to produce something spectacular with the community, so we decided to make an Eday version of a gravity battery—and on those terms, it worked. As we sat around talking into the night, surrounded by absolute darkness except for the lights of neighboring islands and the hostel in the distance, we also agreed that the machine we built was, in some real sense, a social machine. Between Andy's words (and music), Clive's stories, Mike's submarine, Mick's motorcycle, the schoolchildren's drawings, Ivan's joyful exclamations of "happy days!," and Hamish's teenage son melting aluminum in a kitchen pot with a blowtorch to cast spare parts, we concluded that some good had come of the trip. The gravity battery itself, meanwhile, remains on the island, and will hopefully power Andy's Casio keyboard for years to come.[5]

8 DARKNESS

SPITSBERGEN, WINTER 2017

I remember hearing my brother-in-law telling the story of how he and his dad and brother took a cargo ferry from Britain to Iceland in 1973, with the family Range Rover out on the deck, and how they sailed around the Eldfell volcano that was erupting in the Vestmannaeyjar (Westman Islands) south of Iceland, and how the sight of that island of fire in the middle of an endless frigid ocean never left him. It erupted, he said, for the entire first half of the year, from January till June. In fact the first Icelandic emigrants to Canada—to Gimli, Manitoba, or New Iceland as it was known (northern Manitoba being if anything even colder than Iceland)—fled those same volcanic islands almost exactly a century earlier, in 1875. So, when Mick told me this story I thought of my own ancestors on my mother's side, who traveled from Iceland to Canada, with a brief stop in North Dakota, after a volcano wiped out their farmland. They settled in a tiny town even further north than Gimli, a town with the unlikely name of Winnipegosis, which sounds like a deadly Canadian

FIGURE 7 Dogsledding, Svalbard.

disease (death by boredom?). The men all continued to live by fishing, going to Sugar Island every year for winter fishing when the lakes froze over. The women had names like Bergljót, Birta, and Rúni. I still bake a cardamom-flavored *vinarterta* every Christmas in honor of my great aunt Vi, whose name I used to think was short for "Viking."

Now I was crossing the Arctic Circle, further north than Iceland, to the island of Spitsbergen in the Norwegian archipelago of Svalbard. I was looking forward to seeing the island; unfortunately, as I soon discovered, that was not possible. In fact the strangest thing about visiting Svalbard in January, during the polar night, is that you can't see the island at all. I was surprised, when I flew into Longyearbyen, how unsettling an experience that was. Despite the darkness that is drawn like a heavy curtain over the islands from October to February, however, the islanders are lively and happy, even slightly giddy. Children build snow forts under stark electric lights. People ride past on bicycles with studded tires, or on snowmobiles with rifle mounts. Huskies are tied up outside the shops, and you must check your gun at the door. Svalbard has a *Casablanca* meets *Ice Station Zebra* vibe to it. There is a Russian coal mining village nearby. The archipelago lies between Norway and the North Pole, and at 78 degrees north it is home to the world's northernmost settlement. There are northern lights, apparently; although I did not see any natural light, northern or otherwise, during the week I was there. There are also 30 percent more polar bears than humans, according to one statistic I read. The day after we arrived a mother polar bear and her two cubs wandered into town and were gently escorted out again, only to return the following day. Polar bears are white and so is the snow; added to this, everything is in darkness.

The fact that I was there for an island studies conference compounded the sense of absurdity: though I talked about islands all week, I never actually saw the one I was on. It was pitch black when we landed, and blackness followed us from morning till night—times of day that ceased to have any precise meaning. Nearly everyone arrived on the same flight from Tromsø and climbed out of the cabin to go blinking across the runway moonscape. I was ready for the cold (the temperature hovered around zero, often with a strong wind chill), but I was not prepared for the disorienting feeling of being on an island built largely from my own imagination. The conference was called, in all caps, REMOTE.

After a couple of days' conferencing, we went to a dogsledding camp outside Longyearbyen. There were dozens of restless dogs tied up outside their shelters, waiting to run or eat. James and I took a sled out with a team of dogs on a ten-mile run; the younger ones were in the back and the experienced dogs up front. There was no guide—we were driving blind. What I mostly remember, riding through the pitch darkness and hanging on for dear life, is how the dogs were constantly shitting, or trying to shit, especially the younger ones. But the team kept running, so the poor things were pulled along by the other dogs and had to shit while they ran, squatting and shuffling and whining, all the dogs tied together and shitting in turns.

After a long and foul-smelling yet exhilarating journey though darkness with the sled dogs we all went back to a log cabin to get warm and load up on food and alcohol. While we were there the mother bear and her cubs showed up. This news was conveyed to us by a man with a large pistol and a rifle as we warmed ourselves with brandy and coffee. (By law you can only leave the city limits with a high-powered rifle, or a guide who carries one.) I overheard a worrying exchange between the man with the guns and our guide, who also had a rifle but carried it discreetly and put it in a locker at the camp. They stood in the doorway of the cabin, talking and gesturing at the group of us indoors and out into the darkness.

"These people have all signed the waiver."

"Ah good, they have signed the waiver." (The waiver stipulated that if we were eaten by a bear it was not the dogsled company's fault.)

"Look—they're over in Philip's camp, near his tent."

"Is Philip there?"

"Ja, I think so."

"Yesterday they scared them away, but they came right back."

"Ja, they must be pretty hungry. They came up here because of the meat." The man with the guns was talking. He turned to us and pointed to our guide, raising his voice: "So stay with the fucking boss!" Nobody had to be told twice.

The scary thing about bears wandering into settlements—aside from the obvious menace of a large white bear hiding in a blizzard during the polar night—is the suggestion that something is seriously wrong with nature; that hungry bears are a visible sign of climate change. Rising temperatures in the Arctic mean melting sea ice, which in turn makes it harder to find food, and the whole sea ice ecosystem starts to collapse. Studies have suggested that Svalbard is warming at six times the global average.[1] The desperate mother bear—for what bear in its right mind would go near a place full of dozens of barking dogs, shouting humans, and vehicles—was likely trying to find enough food to feed her cubs.

The effect of day after day of total darkness is hard to describe. It wasn't far to reach the end of the road—after which there was only the abyss, like falling off a map. (I thought, "Am I walking into eternity along Spitsbergen strand?") Occasional gale-force winds might whip up unexpectedly, turning a walk to the pub into a blind life-or-death journey in which your colleagues suddenly disappeared and you were walking down an endless icy road, alone. This made one pub on the edge of town feel a bit like Minnie's Haberdashery in the Quentin Tarantino film *The Hateful Eight* (2015). At the same time there is nothing cozier than being indoors, on an island, in midwinter, in the high Arctic. Everyone padded around the corridors and common areas of the hotel in sock feet. We presented our talk in woolly socks. *Gezelig*, as the Dutch say. Similarly *koselig* in Norwegian. *Hygge*, in the now famous Danish expression, or more properly *hyggelig*.

In Madeira this feeling was only possible in the mountains during a storm. In every case the sense of isolation is key.

Polar night puts you in a philosophical frame of mind. Are there really bears out in the darkness? (If so, where exactly?) Is that a mountain or an iceberg? If this is an island, where do the edges lie? To find out you would only need to walk in any one direction long enough, but the threat of bears means you would not get far before someone brought you back to the settlement. Is it true that it is illegal to die in Longyearbyen? (No, but it is impossible to be buried there due to permafrost. Victims of the 1918 flu pandemic are still waiting to decompose—with global warming they may soon get their chance.)

At the conference we met Owe, an ethnologist and musician, a generous soul from the island of Gotland in the Baltic Sea, who showed us his collection of cartoons of desert islands. The presentation topics ranged from medieval Norse-Sámi relations to intercorporeality and islandness to cultural identity and animal husbandry on the Estonian island of Ruhnu (population 97). For our part, James and I spoke about designing gravity batteries and other speculative energy solutions for Madeira—the research we had first taken on tour to Hong Kong the previous year. We sat drinking boxed wine from the Nordpolet late into the night. We discussed more questions: Why is Norwegian cheese brown? What is the best nonlethal defense against polar bears? How long can a person survive without sunlight? Was it really healthy to jump into the snow after a jacuzzi, as we saw people doing at the hotel?

Svalbard may not be a place for normal everyday living (or dying), but we did find promise in the periphery—both in the case of Svalbard and our own remote island. We all survived the polar bears and the polar night, and after almost a week of darkness saw the sun again as we flew over Tromsø. I re-entered the world just in time to watch Donald Trump's "American carnage" inauguration speech on CNN in the airport lounge. Suddenly the remote expanse of Svalbard looked less like a hostile and frozen wasteland, and more like an oasis of calm amid a burgeoning apocalypse. (I'll slip this message into the empty akvavit bottle we used to kill the pain.) At the same time, uncertainty and instability touched this place as everywhere else: shortly after we left Svalbard, the Global Seed Vault's entrance tunnel flooded as the permafrost—which was obviously meant to be permanent, hence the name—started to melt as temperatures soared. No seeds were lost, but it seemed like an ominous sign only a decade after the vault was built as a failsafe deep-freeze to protect the world's crops against harm.[2]

9 CITIES

HONG KONG, SPRING 2016

Because we worked on an island, islandness became the theme of our research, as it did for other researchers in our institute. (Write what you know.) We did fieldwork about islands on our own and other islands. We won island-related grants because we lived on an island—so obviously we were the experts! Our project partners worked at institutes on islands scattered around the world. We met at conferences about islands that were held on islands, talked about islands together, and then wrote about islands afterwards, and this cycle perpetuated itself endlessly. Talk about insular.

Now James and I were speaking at Hong Kong University as part of the latest edition of an island studies conference series: a legitimate yet somehow slightly dubious endeavor that gathered academics on exotic islands around the world to discuss, well, islands, and more importantly to go dogsledding, spelunking, snorkeling, skydiving, or whatever. It was in fact no more dubious than the standard pre-pandemic conference racket—and it was a racket, not least from an environmental perspective—but this series seemed especially emblematic of the excess and touristic emphasis of those heady and unsustainable days of scholarly life in the 2000s and 2010s. We flew Emirates via Dubai and stayed in a hotel in the Western District of Hong Kong Island, where we sat in the elegant lounge drinking whisky each night when we returned from our scheduled itinerary. We ate delicious egg tarts at the Hoover Cake Shop—a Hong Kong staple that the Portuguese claimed to have bestowed upon the former colony via the *pastel de nata*—and dumplings at a steamy hole in the wall somewhere in Kowloon. When I had nothing better to do, I tracked down the shooting locations of my favorite Wong Kar-Wai films like *Fallen Angels* (1995).

Our guide was a jittery raconteur with an unpronounceable Scandinavian surname, who met us on the pier the evening of the first day. He identified himself, secret agent style, by wearing a red handkerchief. He bowed deeply, presenting our boat with a flourish of hands as though he'd just conjured it out of thin air. "All aboard!" he shouted. We walked up the gangplank of the small, semi-covered ferry and sailed to the former fishing community on Lamma Island for dinner. Along the way we made small talk with the other delegates and looked out through fogged windows at the slate gray sea. When I asked our host about his background I was surprised to learn he was American. He told us that he had reinvented himself in northern Europe. (*Mon semblable—mon frère!* I thought, for I had done more or less the same in Portugal.) The exotic surname belonged to his wife. "And now I'm an island guy!" he said. "It never gets dull, unlike most of academia. Islands are the future." His eyes darted around the group. We docked at the smaller island and ate at a touristy seafood restaurant with impressive displays of lobster, abalone, oysters and scallops. After dinner and a few drinks we sailed back to the main island in darkness. James and I spent most of the return journey chatting with John, an amiable British sociologist who specialized in offshoring practices in places like the Cayman Islands.[1] We promised to attend his keynote lecture the next morning despite the early hour.

After John's keynote, which gave me a new perspective on tropical islands I thought of mainly as holiday destinations, but which now appeared to be resting spots for untold and unseen amounts of ill-gotten wealth, it was time to present our research. The talk, which we had put together on the plane, was based on a blog we had started called Crap Futures that had gotten a mention in *Wired* and a few other places. We were concerned with the future, the impact of emerging technologies, and islands. Our guiding lights were J. G. Ballard and Ray Bradbury ("People ask me to predict the Future, when all I want to do is prevent it").[2] The talk featured images of Madeira's vertical terrain, exploratory sketches of speculative designs for the island, and some history of island adaptation and innovation generally—the car made of bamboo from *Gilligan's Island*, for example.

The next day we visited the Design department of Hong Kong Polytechnic, housed in the swirling concrete of Zaha Hadid's Jockey Club Innovation Tower. The tower was an impressive embodiment both of Hong Kong's colonial past and its ambitions as an independent global city. That night I had arranged to meet my uncle at a restaurant in Central. Angus, my mother's older brother, was a tall, handsome ex-air force pilot who had spent much of his civilian life flying for large commercial airlines and living on exotic islands. (All the men on my mother's side were named Angus.) He had remarried and lived a comfortable life in Hong Kong training pilots for Cathay Pacific and putting off retirement. During dinner we talked about my other uncle, who had recently moved back to Vancouver Island after hopping between Fiji, Australia, New Zealand, and Prince Edward Island. Why did our family move so much, I wondered aloud. He shrugged and said we had never really settled anywhere: the three children had moved around Canada as air force brats their whole childhood, and even our Scottish ancestors seemed to have restless feet, coming over originally from Normandy, moving up to Skye, down to Glasgow and Edinburgh, and then across to the New World. I tried to press him on the colonial legacy of Canada and Hong Kong, but I knew he was apolitical by nature. I told him my father's cousin had been a policeman here in Hong Kong, and he nodded and said, "That's right." That's what people like us did—we moved around from island to island in mid-level jobs, rootless and adaptable. I did not feel like a colonizer in Madeira, since I earned a local salary and owned nothing, but I did live on an old English wine family's *quinta*, and I knew how the locals saw people like me: as an unwelcome guest at worst, a necessary evil at best. "It's a good life though," my uncle said. "I'm sure you'd agree." I smiled faintly.

I found Hong Kong in 2016 strangely similar to the Montreal where I lived in the 1990s: boldly cosmopolitan, with an unusual and distinct identity that was resented on the mainland. Now nearly everything we saw and everyone we met there is gone, or immeasurably changed. Our British offshoring friend John died only days after we sat together on the ferry. Zaha Hadid died two weeks later. In 2019 the Polytechnic itself was the scene of some of the most violent and dramatic student demonstrations against mainland control. My uncle left too, and is finally back on Vancouver Island, living up at the northern end where he started decades ago, and where land is still affordable. He retired not long before his 80th birthday—still reluctant, pushed into it by the upheaval of the protests, the pandemic, and the tightening grip of the mainland. (I've tucked this message into a bottle of Blue Girl lager, a German beer brewed in South Korea that is oddly popular in Hong Kong. I hope the improvised wax seal holds.)

10 NOSTALGIA

MADEIRA, 2013–2016

Since I left Madeira it has become increasingly unreal, like an island seen from an airplane window as it disappears into the distance. But it is a magical place. The atmosphere hits you as soon as you step off the plane from Lisbon. The mountains rise up on one side, tall enough to get snow in the winter, and the ocean extends out on the other as far as the eye can see. It is warm and humid; the elderly thrive like orchids. The first time Simone and I landed, Valentina picked us up in her Fiat 500 and we raced along the curving, undulating country road into the hills of Santo da Serra, to the weekend house where our boss had arranged a welcome barbecue with everyone from the institute. There was a husky named Lobo, a rescue and castaway enjoying the cooler atmosphere of the mountains above the clouds. I came to see this Arctic dog, whom I met on several occasions, as a symbol: of the randomness of islanders, and the loneliness of isolation; but also, the uniqueness of islanders, each one a special number in a limited set. Kids were playing football on the wet lawn; it was our first time travelling without our own children. The next three days were a blur of job talks, swimming in the Atlantic, and drinking unimaginable quantities of red wine and *poncha* (aguardente blended with lemon and honey or any of the island's exotic fruits, including resinous *pitanga* and *tomate inglês*). Valentina and I got sea urchin spines in our feet climbing out on the rocks. We moved almost

FIGURE 8 Box elevator, Fajã dos Padres.

immediately and lived on the island for eight strange and wonderful years.

Madeira is a popular tourist destination, advertised as the "pearl of the Atlantic." Like the island of W in Georges Perec's *W or The Memory of Childhood* (1975), it was "entirely uninhabited when it was colonized . . . by the group whose descendants constitute the entire population."[1] (This is a slight exaggeration in Madeira's case, but only slight.) Voltaire believed it to be the most likely location of Atlantis, if such a place existed. Not long after settlement in 1419 it began to serve as a major stopover point for the transatlantic sugar and slave trade, and has been called "one of the birthplaces of capitalism."[2] More recently it has been a refuge and holiday destination for the wealthy. The last emperor of Austria, Charles I, chose the island for his exile after the First World War and died only a few months later, aged 34. He was buried in a chapel above Funchal—except his heart, which is buried in an abbey in Switzerland. The Cuban dictator Fulgencio Batista also chose the island as his place of exile after he was chased from Cuba by Fidel Castro's rebels. Winston Churchill arrived in 1950 to paint watercolors of the bay, drink the fortified wine, and catch butterflies; he was chauffeured around the island in a dove-gray Rolls Royce belonging to the Leacock family. Another future Prime Minister, Margaret Thatcher, stayed in the island's Savoy Hotel on her honeymoon the very next year, but the young Margaret had a less favorable experience than old Winston.

She lamented in a letter to her sister that the island was full of "'tatty' tourists: Jews and the novo [*sic*] riche."[3]

Today the island is mostly known as a place for people who like to walk. Fit retirees from northern Europe flock here year-round to trek the *levadas*—an ancient and seemingly endless network of irrigation channels that crisscross the island. The *levadas* flow between high mountain peaks, through banana and eucalyptus groves, and up on the wild north of the island through the primeval, UNESCO-protected laurel forest that at one time covered much of southern Europe. The trails are mostly flat, making them surprisingly easy to walk: they transport water, and water doesn't like to travel uphill. It's all so beautiful, the visitors say. "Come to walk!" the tourist brochures say. And the flowers—oh, the flowers! That wasn't my island, although I did walk. I spent much of my time walking up and down the hot, vertiginous streets of Funchal, on my way to work across town or doing the school run with my two uniformed kids.

There is so much to tell that I'm going to write it in a style after of one of my favorite memoirs, Joe Brainard's *I Remember* (1975).[4] I hope this doesn't seem like a cheat. Though doomed to incompleteness, Brainard's approach of assembling a thousand fragmentary moments from his life has the virtue of evading false claims to a uniform narrative of one's past and oneself—we are just fragments, and fragments are all we remember, and all we can remember, since every day introduces new events. Each memory fragment resembles a little island in a chain: a never-ending archipelago of memories that make up an identity. I'll put this message into bottle of vintage Sercial that I was saving for a special occasion.

I remember diving into the deep blue ocean to cure a hangover: the sting of salt water, all the wine washed away, cleansed and weightless, the floating platform.

The neighbor dropping by with a husk of tiny bananas, even smaller than the kind the island is known for, yellow and ripe and no bigger than my thumb.

Gathering dozens of banana leaves from the grove next door under cover of darkness, clutching their bulk to my chest and running back to the kitchen where Simone was making tamales with pomegranate salsa for the holidays.

O Castelo dos Hamburguers (Hamburger Castle), a miracle of island ingenuity, serving hamburgers that were fully realized yet unrecognizable, seemingly created based only on the abstract notion with no firsthand knowledge, like Kafka's *Amerika*. Wide and thin, more soy than meat, in a pita pocket with pink sauce, they were best washed down with a tall peanut shake that would induce a mild anaphylactic shock even in the resolutely non-allergic.

Everything at the wrong scale—most things reduced but flora blown up to massive proportions under ideal conditions: the African violet from my grandmother's windowsill thirty feet high in the park; the lavender I once bought in tiny pots now in bushes the length of the *quinta*.

The ants—the ants! Sugar ants that devoured any food left idle for more than ten seconds and infiltrated anything warm in winter, destroying it from the inside: the toaster, the DVD player (twice), the television, the internet router—even the car, which they used as a giant nest, falling away in their hundreds like tiny clowns when you surprised them by opening a door.

Teenagers whacking each other with lengths of sugar cane in the streets.

Bougainvillea, pink, orange, and red, spilling over the stone walls and flooding the garden.

Houses half built, others half decomposed, ruins of past dreams; the house our neighbor built for his son to live in, but he never returned.

The Nun's Valley, hideout from pirates in past centuries, islanders living their whole lives without seeing the sea.

My wife meeting Tony Conrad and his partner on a plane leaving the island, asking where they had been (Ponta do Sol) and where next (Morocco)—and then arranging a miraculous lunch in Marrakesh a week later.

Never knowing anything: the children failing to dress up or bring sweets for feast days; the mystery of municipal laws; information kept close because knowledge is power, an island's best defense against invaders.

The island highway, tunnel after tunnel, the limited network of roads and the feeling of being a rat in a maze. And why does everyone drive so fast? Where are they going?

A taxi driver telling me the odometer on his Mercedes had gone around eight times.

Sand from the Sahara reaching Madeira over hundreds of miles of open sea.

The fragility of connections: stranded for days when a storm swept in, trips cancelled and plans forgotten, whitecaps on the sea, the wind from the north gusting too much for the plane to take off.

The outdoor patio at the airport where you could watch planes perform the tricky bank turn and land on the short runway, elongated since one unlucky plane fell off the end, crashing onto the beach and exploding into flames.

The sound of wild clapping every time the plane landed and being told by the flight attendant as I disembarked: "Enjoy your holiday!"

The other kids calling our son *lourinho* ("blondie") and he and his sister being exempted from confession, the unexpected lesson of difference.

The pale green abandoned school, with bananas growing inside its gates and intricate pebble mosaics covering the grounds.

The tiny dark shop across from the school where we used to buy wine, the empty shop with its ghostly shopkeeper that was once filled with the cries of children.

Thunderstorms followed by the heaviest rain, always just before dawn.

Thinking that if the sun killed my dad, is moving to a subtropical island really a good idea?

Our first rented house, plagued by termites and mosquitos and cockroaches, dilapidated and cavernous and right in the dengue zone, yet somehow perfect.

Boxes from the mainland arriving months late, with all our worldly possessions that in the meantime we had done very well without.

Living in paradise on a string of short-term contracts. Wondering if we would ever make it off the island.

Working out ideas with James around the fire with a glass of wine, because he said (and he was right) that it's the best way to get things done.

The art of making do, island *jugaad*, like shoving slices of cork into the gaps of the window frames to stop them rattling.

The road in front of our door so steep that it turned into a rushing river during heavy rain and being advised by the neighbor not to open the door until the rain stopped.

The name John Dos Passos on a plaque in the village of Ponta do Sol, where Manoel dos Passos fled to America after a knife fight, and young Dos visited with E. E. Cummings after the war.

The story of Hemingway, frenemy of Dos Passos, stopping off en route to his *finca* in Cuba in 1954 and refusing to get off the ship, floating instead in the Bay of Funchal while his wife Mary bought lace, drank Madeira wine, and rode the street toboggan down from Monte in the company of the ship's captain.[5]

The aristocracy at island scale, famous families who drove their kids to school in old beaters and went out for pizza and laughed and drank too much wine just like us.

How after two centuries on the island they were still treated like foreigners.

Thinking: what chance do we have?

Spending afternoons in a dingy pub called the Number Two, watching tourists order tankards while I nursed a small beer and waited for the kids to finish lessons at the old conservatory.

What the island doctor said in *The Mahé Circle*: "to live here, you don't want to be ambitious."[6]

When the institute for a time had two warring fathers, and a strange civil war consumed and nearly destroyed it; but the island was immutable and soon regained its equilibrium.

Our young colleague, whose tales of wild sexual adventures and ayahuasca journeys shocked and delighted us, and how worried we were when he fell into a hole and lost a tooth.

Walking down the side of a volcano on my daily commute, breaking into a run because the incline was so steep.

Standing in a ditch waiting for the cars to pass on the narrow road and thinking of Beckett.

Looking up and noticing a visiting researcher from MIT with a beard and glasses standing in the ditch opposite.

Getting mad and knocking the wing mirror off a passing car with my daughter's violin case, the fourth car through the crosswalk and he hadn't even slowed down.

A loud crack as it went sailing through the air and landed, skidding briefly along the road, and the Fiat Panda screeching to a halt, the driver in cut-offs and a Cristiano Ronaldo haircut.

How he called me son of a whore and my daughter started to cry, but he saw my willowy rage as I held her hand, and he shook the broken mirror at me and sped off.

How in summer the teenagers became amphibious and lived as much in the sea as out of it, diving off the rocks and stretching out to bask like lizards on the concrete dock.

The feeling of swimming, surrendering yourself to the irresistible pull of the ocean, then climbing the stairs to the bar at Barreirinha for a globular gin and tonic.

The islanders who lurked resentfully in the shadows of the Old Town, turning their backs as if to say, who loves the sun?

The packs of stray dogs of all breeds, big and small together, stopping to rest on the hills or lying undisturbed in the middle of the road while cars drove slowly around them.

The night market in December with stalls selling fresh *bolo do caco com chouriço* and glasses of *poncha*.

Seeing a Christmas tree made of salt cod in the supermarket, festooned with onions and garlic, the salt glistening like frost under fluorescent lights.

The fire that burned from the mountains right down to the sea, the city at night dappled with orange and the smell of heavy smoke, being trapped between mountains and sea anytime disaster struck, whether fire, flood, earthquake, or virus.

Camping with the kids on the north of the island, the whaling lookout and the crashing waves, walking through endless tunnels to avoid falling rocks, and the big red moustache I grew for the occasion.

Madeirans using the sea as an extension of the island, for business and gossip, and learning to use it that way too.

The Principality of Pontinha, a micronation in the port founded by "Prince" Renato de Barros, a local art teacher, who bought the islet from one of the old wine families for a small sum and powers it with wind and solar.

Eric's forty-eight-hour visit, packing in a lecture, a swim during a winter storm, and a *levada* walk with a chain-smoking guide who yelled at us to stay with the group as we dashed on ahead, feckless as when we were teenagers.

The crudely welded metal box, size and shape of an old telephone booth, winched half a mile up the sheer side of a cliff over the beach at Fajã dos Padres.

The kids wanting to ride in the metal box, saying yes and then watching in horror as it moved slowly upwards at a dramatic angle suspended on a long, thin wire, the latest in a history of island transport innovations from elevators and gondola lifts to cog railways and oxen hammocks.

Our kids, now grown and serious, so beautiful and innocent on that far off island.

The flowers and sun and sea, as well as the traffic and terrible isolation.

The winter just before Hong Kong: seeing you standing at our door with your new partner, tall and slim with short blonde hair, not to visit us, as it happened, but to dance, because she was a dancer, and now you were too.

How you were improbably lithe for a middle-aged man with a beard.

Sitting by the fire with you at James's on the *quinta*, drinking wine and talking in the dark as the rats scurried along the trellis overhead in the yellowish light.

How you noticed the drawing of Brendan Behan in our kitchen, and worried about your father who was almost gone.

The abandoned house in the Old Town, the floorboards missing and the roof falling in, dancing and drawing on butcher's paper with the crowd following transfixed, one of those rare island moments when anything can happen, being so far from everything.

Joking: "dance like no one is watching"—to which you answered: "because no one is."

That warm night in the mid-Atlantic, sitting out behind an abandoned house, perched on the rim of an old well, with the shadows of palm trees and the ghosts of sugar factory workers, watching you drop small stones in one by one as we drank and talked. Plip. Plop. Plip. And then you were gone.

The very last time I stood beside you, in Lisbon not long after: you wrote your name in pencil on the restaurant waiting list right below Jesus. You picked up a bathroom sink on the street and wanted to take it home to Canada. We drank wine in the square with the old men and hoped someday we would be as happy as they were.

Simone documenting the out-of-time strangeness of island life in a series of black and white photographs she called Tar Island,[7] a place full of broken concrete and stray dogs and terrible beauty, where nothing moved in the heat and everyone was stranded and waiting.

The beauty and ugliness of life, the pleasure and pain, all compressed on this dying volcano surrounded by an oceanic trench two miles deep.

11 SECRETS

IRELAND, 1998–2000

I once told a friend from Belfast that my Irish grandfather used to shave twice a day—a habit that always struck me as excessive. He asked with a dry smile: "Your grandfather wouldn't have been a Protestant man, by any chance?" I said that yes, he was a Protestant man. He and his many brothers—there were ten siblings in all, Protestantism notwithstanding—left for Canada beginning in about 1910. That part of the story is not about islands, except the need to escape them. However, my grandfather's oft-repeated line, "Whatever you say, say nothing," strikes me as a very typical thing for an islander to say. He said the words with a mix of residual trauma from his difficult upbringing on the wrong side of the border, and a touch of ironic humor from the safe distance of his adopted home on Vancouver Island. Ireland has its share of dark secrets, like the Magdalene Laundries that operated from the eighteenth century until the late 1990s, with horrific burial sites recently unearthed that echo those of the Canadian Indian residential school system (as it was known) also run by the Catholic Church. My father used to talk about Ireland as a grim place, a terrible backwater that his father had fortunately escaped, but to me it was infused with romanticism despite its flaws. "Island of saints and sages," as Joyce called it—notably from exile in Trieste.

My first term at Trinity College, after I fled grad school in the American Midwest, had its ups and downs. Academically speaking, it was mostly spent in smoky Northside pubs, listening to moody Irish ballads and falling prey to infatuations. I was attending too few lectures, and fewer sober; yet somehow I soaked it all up. I read voraciously, either in my green leather nook at the back of the Stag's Head or in bed all day in my rented room while I nursed a hangover, like the feckless student narrator in Flann O'Brien's *At Swim-Two-Birds* (1939). ("I know the studying you do in your bedroom, said my uncle.")[1] Not only Flann but Wilde and Joyce and Behan and O'Casey and Yeats, Elizabeth Bowen and Edna O'Brien, Jennifer Johnston and Eavan Boland, Brendan Kennelly (one of my teachers), the playwright Marina Carr (another teacher, who poured out glasses of wine for us in her rooms), and Seamus Heaney (who gave a reading across Nassau Street in Fred Hanna's bookshop just before it closed). I waded through the mystical mire of Yeats's *A Vision* (1925) and read the notorious *Black Diaries* of Roger Casement, the colonial civil servant turned human rights activist and gay Irish revolutionary, who was caught running guns from Germany to Ireland and executed by the English for treason in 1916. My blood ran black and white, and my eyes puffed up from the strain of reading fifteen hours a day. But the island felt like paradise, and I found a small community of likeminded obsessives. When I first moved to Dublin I lived in a room in Sandymount, near the beach Stephen Dedalus paces in the "Proteus" episode of Joyce's *Ulysses* (1922), testing the phenomenological reality of the world around him, stepping on shells ("Crush, crack, crick, crick"), smelling the briny air, asking himself: "Am I walking into eternity along Sandymount strand?"[2] Unlike Stephen, who reluctantly returns to Ireland after receiving a telegram ("Mother dying come home father"), Joyce became a permanent exile on the Continent: for the rest of his life, moving from Trieste to Rome to Zurich to Paris, he only ever wrote about the real and imaginary characters on his island. To my Joycean delight, my girlfriend at the time lived on Usher's Island—once an actual island in the Liffey, now a quay—where Joyce set his long short story "The Dead" (1914). (The climactic scene of that story, the memory of the lovesick Michael Furey shivering beneath Gretta's window, takes place on yet another river island: Nuns' Island, located in the River Corrib in the center of Galway.) I wrote my dissertation on Joyce and Walter Benjamin, spending hours trying to find out if, for example, the two had met in Paris, perhaps at Sylvia Beach's bookshop? I found no evidence of a meeting, but such were my thoughts at the time.

That spring—needing a break from my studies, with St. Patrick's Day around the corner, and feeling exceptionally

Irish with a new ancestral passport—I decided to visit the family farm. Aunt Edna, my grandfather's sister-in-law, a woman from Sligo whom I'd never met, was now the sole occupant. I wrote her a letter, and a week later she wrote back, inviting me to come up for the long weekend. On Friday I walked down to the Bus Éireann station and caught a bus to Belfast. I got off a few hours later in the small town of Clones. I found the place cold and gray: no wonder they left, I thought. But suddenly there waiting for me was Edna, a robust widow in her eighties with thick glasses and gumboots. She said hello without offering a hug and drove us in a battered Mercedes back to the farm at Smithboro, County Monaghan.

My grandfather was born in 1913, a year before the Home Rule Bill was passed by the British parliament but then delayed. He was at the younger end of the siblings, so he never met his oldest brother Robert, who had already left to settle in Canada, as most of the siblings would, only to be sent back to Europe to die in Vimy, France in 1917. Another brother, Ernest, born in 1899, was a professional football player. When I met Uncle Ernie as a child I was impressed by his sporting exploits, but more impressed by his age. Imagine knowing someone born in the nineteenth century!

I recently discovered my grandfather's private memoir in a box of old things. In it he describes what a public education on the colonized island of his first eight years looked like:

In geography we memorised all the counties of England and all the English rivers and lakes. In history we memorised the names of all the English kings from William the Norman and William his son right down to George V, and we learned how the Battle of the Boyne was won. It was all very confusing at the time.[3]

After 1921 things changed dramatically, when he found himself living in the brand new Republic of Ireland. As a Protestant he was on the "wrong" side of the border, which was less than two miles from his house. Northern Protestants were British to the Irish, but Irish to the British. This made things awkward: "while the ten Hanna children were growing up it was in a political atmosphere of Unionist squared off against Sinn Fein with frequent violent clashes," he wrote. A small island had been cut in two. I read with shame and horror—being more sympathetic to the Republican than the Unionist cause—that my great-grandfather Thomas Hanna had been, in my grandfather's words, "for decades the leader of the Orange and Black Institutions in the County of Monaghan and as such a leader of all Unionist activity in the County and beyond." This made things even more awkward for the Hannas, and it is no wonder most of them left.

As I was beginning to learn, however, Irish history is complicated. The sectarianism of the twentieth century was only the latest round in a long match. An earlier Thomas Hanna, my grandfather's great-great-great-grandfather (or thereabouts), from the same part of Ireland, was tried for treason as a local leader of the United Irishmen after the Rebellion of 1798, when Catholics and Protestants joined forces in an attempt to expel the English. This was something I could be proud of—along with the illustrious list of Irish authors and revolutionaries who happened to be Protestant, from the male-dominated pantheon of writers like Swift, Yeats, Wilde, Shaw, Beckett, Synge; to Lady Gregory, Maud Gonne, Jane Wilde (Oscar's mother, the nationalist poet known as Speranza), the poet and suffragist Eva Gore-Booth, and her older sister the revolutionary hero Constance Markievicz.

When you emigrate to the New World no one really cares which side you were on. My grandfather became a moderate Irish nationalist, shedding the narrow sectarian views he grew up with for a more enlightened perspective. He later served as a Liberal member of parliament in the traditionally conservative province of Alberta and was deeply read in history and literature. He liked to repeat his favorite line of Brendan Behan's: "I'm not a British subject, I'm a British object!" (This is according to my aunt Nora, who remembered Behan's voice echoing through the house from recordings of his songs and readings. I have no idea of the actual providence of the quote.) Loyalist by birth but not by belief, he was disappointed that he wouldn't see a united Ireland in his lifetime, even though partition was meant to be a temporary measure. When I was eight or nine he talked me through the horrors of the Troubles—at that time headlines in North America were dominated by news of the hunger strikes in the Maze prison in Northern Ireland where Bobby Sands and nine others would eventually starve to death. He spoke of the coldness and cruelty of Margaret Thatcher looking on from her island, but also the curse and stupidity of sectarian violence. Like Yeats (without the later quasi-fascist leanings) he was both proud and appalled at what his country, a small island of a few million people, was capable of doing as the world watched.

When he arrived in Canada he found work as a teacher in the town of Hanna, Alberta (no relation), hundreds of miles from the sea. But during the Second World War he was stationed at the North end of Vancouver Island—the first of my family to set foot there—as a Royal Canadian Air Force officer, patrolling for Japanese submarines in the Pacific. They used seaplanes armed with a machine gun that fired through the propeller. He recalled that the north of the island was still mainly inhabited by indigenous people at the time. In 1944 he was abruptly sent to Britain for the remainder of the war, and he and my grandmother sold their log cabin near the seaplane base for four hundred dollars including all its contents. While stationed in Britain he became friends with a French-Canadian war correspondent named René Lévesque, who would later become the Premier of Quebec and leader of the sovereignty movement there and the first referendum on independence. My grandfather went home to visit his mother at the farm in Smithboro during leave in 1945. (My other grandfather was also stationed in Britain during the war, where he served as a motorcycle dispatch rider, navigating the narrow country lanes without the use of lights.)

By the time I arrived in Monaghan I knew not to expect much of the legendary family farm, and in this lack of expectation I was not disappointed. There had once been a larger house, Edna told me, but it had been torn down in the sixties. In its place was a small and sensible two-story stucco house. There were a few crumbling outbuildings, several sheep out front that Edna called "pets," and some enclosures behind the house which held five bulls and two or three horses. Although she lived alone there were a couple of local men who worked the farm, and her niece Ruth, my father's cousin, stopped by most days. Inside the house was a mix of the very old—somber furniture that, having survived the journey to that remote house, would never leave—and the strikingly new, including a huge television positioned directly opposite a sleek black leather lounger.

On Friday night Edna served fish fingers and boiled potatoes for dinner. Since it was just the two of us we ate in the kitchen, and afterwards retired to the living room. There we sat, Edna in her lounger and I on the lace-covered sofa, watching John Wayne in *The Quiet Man* (1952) and saying very little. I was beginning to realize that, unlike the Dubliners I had met, Edna was a woman of few words. (I remember noticing that John Wayne's trousers were pulled up higher than any trousers I had ever seen.) Eventually during a long commercial break we started to talk. She told me the history of my family, once prosperous "gentlemen farmers" now reduced by emigration and economic crisis to this lonely widow living in a few rooms of a modest country house. I asked Edna if she had seen any famous productions by Yeats or Shaw at the Abbey or the Gate. She indulged me, saying she had definitely seen something scandalous by Shaw.

She also told me a story involving Oscar Wilde's two half-sisters, Mary and Emily. I knew that Oscar's father, William Wilde, was a notorious philanderer (or worse) who had children hidden away in houses up and down the country. Two of these children had lived on a nearby estate. They had died together, Edna told me, in a tragic fire in the house right next to ours, shortly after Oscar's seventeenth birthday— though it was unclear whether he even knew of their existence. On October 31, 1871, during the last dance of a country ball, the hem of Emily's crinoline gown had suddenly burst into flames. Crinoline was notoriously flammable, so much so that death by crinoline fire was not uncommon. Mary tried to rescue her, but she was also wearing a crinoline gown and soon caught fire as well, and both sisters suffered mortal burns. William Wilde, Edna told me with a sideways glance, had been spotted at the graveside in the weeks after the funeral, wailing openly in his grief. He never recovered, she said. Was she embellishing a bit perhaps? But no: he died a few years later, by all accounts a broken man, not unlike Oscar in Paris after his release from prison.

The story came up completely by accident. Not long after I arrived, I noticed a copy of Richard Ellmann's biography of Wilde sitting primly on a doily covered china cabinet.

Ellmann, the American son of a Jewish Romanian father and a Ukrainian mother, was Goldsmiths' Professor of English Literature at Oxford University from 1970 to 1985. (He also passed through Trinity College, Dublin.) I had taken down several of Ellmann's books from the stacks at Trinity library during my first two terms. I guessed he might have written about Wilde reluctantly, being unsure what to do with him: Wilde was modern, but not exactly a modernist; he was gay, which Ellmann seemed to have difficulty talking about; and unlike Joyce or Yeats, he seemed to have left his Irishness behind when he left Ireland. Ellmann struggled with the biography through the last two decades of his life. When he died in 1987, the same year the book was published, *Oscar Wilde* was posthumously awarded both the Pulitzer Prize and the National Book Critics Circle Award. (The book was later used as the basis for the biopic with Stephen Fry giving an uncanny performance as Oscar.)

I asked my aunt about it. Ellmann had come to Ireland to research the book, she said, and one of his stops was in Monaghan to investigate the story of Mary and Emily. As Edna told it, he had lain in wait outside the local church on Sunday, and when the congregation emerged Ellmann asked if anyone knew the story of the sisters' deaths. Someone pointed to my great uncle and said, "Ask Mr. Hanna, he'll know." When the book came out a signed copy was sent in thanks.

The story of Wilde's sisters that my uncle told Ellmann is a sensational and tragic one. Yeats's father recalled the sisters' death in a letter in 1921, so the story was probably familiar to the small world of Dublin society. At the time of their death Mary and Emily were wards—like Cecily Cardew in *The Importance of Being Earnest*—of William Wilde's eldest brother, the alliteratively named Reverend Ralph. The Reverend Ralph, who christened Oscar, was rector of St. Molua's, Drumsnat: the parish church that my family attended in Monaghan. The neighbor's house, where the party took place, belonged to a local bank manager named Andrew Reid. Reid was the man who had taken the last dance with Emily and then tried in vain to save the sisters when their dresses caught fire.

The night of October 31 there was a party to celebrate All Hallows' Eve, or Samhain in Ireland. It was attended by the well-to-do families in the area, from neighboring estates like ours. (I asked whether it was likely that anyone from our family had been present, but Edna just shrugged.) There was plenty of alcohol, and the party went on late into the night. Some accounts describe snow on the ground: Reid is said to have rushed Emily outside and rolled her in the snow to put out the flames, while Mary ran around screaming frantically until she collapsed. There is no mention of snow in the official inquiry; but then again the inquiry also gives the family name not as Wilde but "Wylie." The aftermath of the tragedy was, if possible, even more gruesome than the terrible accident itself. The sisters remained in the house, where they were treated for the severe burns they had both suffered. To die on Halloween night would have been merciful: instead they lingered on for days and weeks at Drumaconnor. Mary, the younger sister who had tried to help, died first, on November 9. Her death was kept a secret from Emily, who was also near death, to spare her the shock; nevertheless, she died three weeks after the accident, on November 21.

The deaths were intentionally hushed up, and details kept to a minimum to avoid scandal. The story ends in the tiny churchyard of St. Molua's, Drumsnat parish, two miles from Smithboro, where I drove with Edna that Sunday to visit the graves of our ancestors before catching the bus back to Dublin. In the car on the way Edna told me the local story of the "woman in black"—thought to be the girls' mother— who visited the graves regularly for twenty years after the tragedy. Oscar Wilde also used to tell the story of a woman in black. Wilde recalled an unknown woman's visits to his house during his father's last illness. The woman would come into the house and kneel by William's sickbed, while Oscar's mother stood by watching.[4]

We entered the churchyard through the wrought iron gate and explored separately in silence. Edna's hands were clasped behind her broad back, her head bowed. Right away I noticed that among the names on gravestones that I could read, at least half bore my family name. There was Thomas Hanna, and Stephen, who died in 1835, and his brother James, and their sister, whose name I couldn't read. Edna pointed out the grave of another great aunt, Amy Elizabeth, who my sister is named after. I knelt in the grass and took some pictures. The grave bearing the names of Mary and Emily was there too, and I photographed it. In contrast to their younger brother, whose famous tomb I had seen in Père Lachaise cemetery, the sisters were all but anonymous, their gravestone untended and overgrown and lost to time.

Years later I went back to Smithboro and the churchyard of St. Molua's. Things had improved. The Oscar Wilde Society had erected a new monument beside the old one to mark the Wilde sisters' final resting place. The simple stone read:

In Memory of
Two loving and beloved Sisters
EMILY WILDE aged 24
and
MARY WILDE aged 22
who lost their lives by accident
in this parish in Nov 1871.

They were lovely and pleasant in
their lives and in their death they
were not divided
(II Samuel Chap. I, v 23)

I'll tuck this message into a bottle of Bushmills—neither the first nor the last.

12 PLEASURES

IBIZA (EIVISSA), SUMMER 1997

Ibiza was great in 1997. Apparently, it was even better in 1987, the year Freddie Mercury celebrated his 41st birthday at Pikes Hotel with 700 friends, accompanied by as many bottles of champagne, cocaine on silver trays, and fireworks. The guest list included Tony Curtis, Naomi Campbell, and Jean-Claude Van Damme. (The hotel was also the setting of Wham!'s breakout single "Club Tropicana.") It might still be great: closer to the mainland in the Balearic archipelago than Majorca and Menorca, Ibiza has experienced wave after wave of paradise-seeking hedonists since it first began to attract European tourists in the 1930s. Islands represent isolation and insularity, but an island like Ibiza is also an international zone, a place of convergence and chance encounters. As Lawrence Durrell wrote of Cyprus: "that is what islands are for; they are places where different destinies can meet and intersect in the full isolation of time."[1]

I took the ferry from Dénia with three Americans I'd met in Prague: two girls from Chico State and a guy from New Jersey who bore a passing resemblance to Jon Bon Jovi. Our first clubbing adventure was a near disaster. Non Bon Jovi stole a whole book of tickets to Pacha from the hotel desk, tickets worth fifty dollars apiece even back then, and told us we could go clubbing every night for free. I had my doubts about the Pacha scene, but as long as it was free I was willing. That night after a late dinner and drinks we used four of the stolen tickets to get in, thinking nothing of it. We were just starting to sweat when three security guards grabbed us and took us to a room for questioning. Suddenly the island changed from paradise to prison. They subjected each of us to a light interrogation, but fortunately we all told the same lie—that we had been sold the tickets by a random guy in town—and to our surprise they believed us. They asked us to go with them to look for the guy, but we said we just wanted to dance, and again to our surprise they let us, even buying us a round of drinks. (I felt some guilt until I remembered how much the tickets cost.) The island switched back from prison to paradise, and we raved with abandon through the Ibizan night and woke up on an empty beach. Once we had thrown away the cursed book of tickets, the rest of the trip was fine: we spent our time drinking, dancing, fucking, and riding like maniacs back and forth across the island on rented scooters. We sailed to the mainland a week later, our brains pleasantly scrambled. All in all it was a fairly typical Ibizan experience.

The philosopher Walter Benjamin found a similar paradise, minus Pacha, when he arrived in Ibiza in 1932. Ibiza represented the opposite of modernity: an island outside time altogether, where a person could get some writing done. Benjamin's life on Ibiza in the 1930s, and earlier on Capri in the 1920s (where he met and fell in love with Asja Lacis), had everything Berlin lacked: sunshine, beauty, freedom, romance, open spaces, a simple daily life. He started each day with a swim in the ocean, he read detective novels by Simenon, and he became more in touch with nature (he noted that there were seventeen varieties of fig on the island). At the same time, Ibiza and especially Capri were cosmopolitan meeting places: "However isolated Capri might have seemed, the hum and buzz of the 'big world' often intruded on the island." He enjoyed the company of Bertolt Brecht and Ernst Bloch; he had tea with the Italian Futurists. He loved the island so much he "considered in all seriousness the possibility of living in one of its large caves."[2] What was lacking on both islands, as was so often the case, was a means to live, and this became a more dire problem in Ibiza. As Benjamin himself wrote: "there are places where I can earn a minimal amount, and places where I can subsist on a minimal amount, but nowhere in the world where these two conditions coincide."[3]

There is a photograph from May 1933 of Benjamin with Jean Selz, a young French art critic, and two others on a boat in San Antonio Bay, where Benjamin looks tanned and pleasantly stoned. He had been experimenting for several years with hashish, as well as mescaline and opium (which he smoked with Selz in Ibiza). On a personal level, however, things were not going so well: he was increasingly broke, living mostly off handouts, and the islanders took to calling him "*el miserable*." Feeling his age, rejected in love, and watching the Nazis rise to power in Berlin in 1932 he decided to commit suicide once he reached the mainland, going so far as to draw up a will and write farewell letters to friends from Nice. Fortunately he changed his mind, and the next year he returned to the island. Even more seriously, however, Ibiza now became a refuge: when Benjamin returned to the island for five months in 1933 he was essentially placeless; the Nazis had taken control and his books were among those being burned in Berlin and other cities across Germany. Although he was relatively safe in Ibiza, he suffered in poverty as his paid work dried up, and he could not escape reminders of the rise of fascism in Europe—the Spanish dictator Franco visited the island in 1933, just as Mussolini had come to Capri during his stay in 1924.[4]

Later I discovered the postcards Benjamin sent from Ibiza in 1932 and 1933. He sent a postcard of a windmill, a view of town from a distance with sheep grazing in the foreground, and a view of Es Vedra island, an uninhabited rock sticking out of the sea off Ibiza's southwest coast, much like the ones I would later know in the Bay of Funchal. He sent postcards from Mallorca too, mostly views of Palma.[5] Benjamin loved to send and receive and collect postcards of all kinds—now, though not then, a melancholy and dying art if ever there was one. I remember writing postcards from the beach in Ibiza and sending them back to friends in Canada. Meanwhile my girlfriend at the time, whom I had met just after my father's death, sent dozens of postcards *post restante* to Pamplona, Spain, from Montreal where she had moved after Vancouver Island. I stopped in Pamplona for the running of the bulls after leaving Ibiza, but that whole week was a blur; I don't think the post office was even open during the fiesta. I never found out what happened to the postcards. Though I visited Montreal the next winter during the historic ice storm, our relationship never quite recovered.

The Irish playwright and ex-IRA volunteer Brendan Behan discovered Ibiza in the 1950s, when it was still a relatively quiet bohemian enclave. Ibiza soon became Behan's favorite getaway, although he would live to enjoy it only a few more years before he drank himself to death aged 41, at the height of his fame. As it happened, J's father Jim met Behan on the island in 1958 and made a simple line drawing of him—which, being the generous person that Jim was, he gave to me when he heard I was interested in Irish literature. (In fact Jean Selz made a line drawing of Benjamin in Ibiza that is not so different from Jim's drawing of Behan, capturing a portly and somewhat ruminative figure.) It was most likely during the winter, in January, when Behan was working on an early version of his most famous play, *The Hostage*, which would debut in Dublin that summer before it moved to Joan Littlewood's Theatre Workshop in London and eventually to Broadway: an all-island genesis,

FIGURE 9 Jim's sketch of Brendan Behan. Courtesy of author.

in fact, from Ibiza to Ireland to Britain to Manhattan. Behan
would have been 100 at the time of this writing, and Jim would
have been 90; Jim was 25 when he met Behan, and Behan was
35. The bit of ink on paper, a momentary sketch done sixty-
five years ago and ripped from a sketchbook, is now hanging
on my wall. The subject, the artist, and his son—my friend—
are all gone. I once asked Jim what Behan was like in person,

what he remembered about the famously gregarious writer. He said: "Drunk. He was drunk most of the time, unfortunately, and he was drunk when I drew him. But he seemed like a nice fellow." (For a detox, and in the spirit of the rave, I'll put this message into a water bottle.)

13 CROSSINGS

VANCOUVER ISLAND — HAWAII, 1996–1997

Sometime during my last year or two of college my father took an unexpected trip to Hawaii. This was highly unusual for a man who never went on exotic vacations and rarely ate in restaurants or bought clothes, although he did like Hawaiian shirts and cocktails. He had never talked about going there before. All I knew about Hawaii was *Magnum PI* and a song by Vancouver punk band the Young Canadians ("Let's go to fucking Hawaii / Get drunk in the sun"). Suddenly he was living the good life, telling me in a postcard about being up in a glider, sailing silently over pineapple plantations.

In hindsight, I think he may have already known something was wrong. A week after I graduated he was diagnosed, following a seizure, with terminal brain cancer. My lease was up anyway, so I quit my pizza job and packed my books and moved from Montreal back to Vancouver Island.

He lived through the summer and died in September. Shortly before he died, he took me on an awful but necessary outing, declaring one morning: "I want to say goodbye to the neighbors." We walked up to each neighbor's house, ringing the bell, introducing me and then breaking the news, saying a preemptive goodbye, watching as the faces of these shocked strangers twisted and fell, as the door closed awkwardly—all in a short visit, repeated at the next house. It was, for me as a young person, almost literally mortifying, but it was completely in character for my father. He looked so different by then, with his shaved head instead of black hair with graying temples; stranger still, he'd shaved the red mustache he was known for—until then I had never seen his upper lip. His death, at least, was peaceful: the room became a quiet island of mourning, in a house that belonged to his third wife, married in a ceremony just days before; all his family gathered around, holding him, then holding his body.

My dad was a psychologist. When he died one of his friends told me it was ironic that he died of brain cancer because he was so smart, like the two were connected. The modest postwar bungalow he bought after his second divorce was nearly empty. He made a conscious effort to square things away in the weeks between his diagnosis that summer and his death in the autumn, but the house was already sparse: echoing rooms with hardwood floors, the basement smooth concrete, all swept clean and free of clutter. It occurs to me that maybe that was why he wanted to move to the island in the first place after the divorce: for a clean start, a clean slate. My only job for the first few weeks, aside from making sure my two teenage sisters didn't tear the place apart when they came to visit, was to dispose of my dad's personal effects. It took about an hour. As a last chore he also named me executor of his will. A lawyer called Dave took care of most of it, sending me papers with sticky cellophane arrows pointing out where to sign.

Killed by a mole, I thought—and so quickly! What hope did any of us have, what certainty could be found in life? None at all—that much was clear to me now. I sold the basic furnishings in one lot, which left only a small, motley collection of odds and ends. I found, for example, an unopened pair of Hanes Y-front briefs. My dad grew up in the 1960s and was a lifelong nudist; he never, as far as I knew, wore a pair of underpants. These must have been a gift from a girlfriend who hoped (and failed) to change him. I also found a stack of soft-core *Oui* magazines from the seventies and eighties. The lesson here was that death is not always as noble in real life as it is in fiction. I was young, and the internet was still dial-up, so I kept them. The porn stash of the father becomes the porn stash of the son. There were some hot tub cleaning supplies. When he was sick, my dad admitted to me from his hospice bed that he had considered ending his life in his favorite place, the hot tub in his backyard. He said he planned to do it with pills and vodka, but he only fell asleep for a while. This was all very sobering news for a twenty-four-year-old.

Grief is a lonely island. After he died, I used to walk his beagle Darwin through the cemetery two blocks from the house and down to the sea. I felt bereft, at sea myself, a big hole cut in the middle of life. People recognized the dog but didn't know me. I was living in my dad's house, walking his dog, looking after my sisters, then fourteen and sixteen, inasmuch as they needed looking after. During that time we went a bit feral: we expressed our grief through partying. Friends came and stayed for a week or two. We tried not to damage the house. Then the house sold; my sisters took Darwin to their mother's house in the country, and I needed a new place to live.

I found a roommate, a guy I knew from Montreal. He was sweet and wise beyond his years; our personalities matched well enough that we managed to share a tiny studio apartment. (The owners likely rented it to us thinking we were a couple.) We made separate "rooms" out of bookshelves; the balcony where I went out to smoke was bigger than the studio itself. At night we would walk over to Dairy Queen, or around the corner to Wah Lai Yuen bakery for red bean buns or almond cookies, or to the Fan Tan Café after midnight where I used to hang out as a teenager. The bright-eyed hippie girls who worked in the cafe on the ground floor of our building gave us free coffee and scones.

I met someone at the record store where I worked and she spent a lot of time at our place. She had never met my dad but must have felt like she knew him since I talked about him constantly. She and my roommate were both

Chinese-Canadian, and by a strange coincidence we lived in Chinatown, which on the island was only about two square blocks. (Neither of them spoke Chinese, any more than I spoke Irish.) Despite these friends Vancouver Island now represented only grief and the past, and I wanted to leave as soon as possible. At the same time, when the worst happens it can energize you in a strange way. I knew this already from a friend in Montreal whose dad had been killed in a car accident which he and his brother had survived. Since then he was completely fearless. During that year I felt invincible, even more than people in their twenties normally do. I wasn't intimidated by anything.

A funny thing happened with the roommate. When we lived together he was studying for a master's degree at the small island university. We both left the next summer. I took a break from island life and went to grad school in Iowa, briefly—the Midwest with its cornfields and oceanless horizons was beautiful in many ways but I felt profoundly out of place in the vast expanse of the continental US. I soon got the kind of island fever that draws you *back* to islands and left for Dublin. Meanwhile the roommate started a PhD at the University of Hawaii. We lost touch for a while, both of us busy in our late twenties and early thirties on our far-flung islands, but later I heard that he had met someone in Hawaii and got married and they had a kid; my story was similar, and I was happy for both of us. After the cold of Montreal and the rain of Vancouver Island, Hawaii seemed like a place where he could flourish.

The roommate's wife had a famous brother. That wasn't always the case—when they met she had a normal brother, a lawyer who lived in Chicago. Everything was normal. But then her brother decided to run for President, and suddenly nothing was normal. I got in touch with him again around this time, and we exchanged messages about how weird it was—he was on the news! That was January 2008. In a further and even more unlikely twist, the brother actually *became* President. For the next eight years the roommate was part of the presidential entourage. I noticed on Instagram that he had started wearing sharp *Reservoir Dogs*-style suits and sunglasses, and with his long black hair he looked like a fashionable assassin. Sometimes he texted me about the strange life he led, like being mentioned on Stephen Colbert's show. As he was becoming part of the entourage, and the global economy was collapsing, Simone and I moved with our small children to Portugal. Our years in Lisbon would be defined by austerity, and eventually this would lead us to the island of Madeira. (This message goes into a bottle of Blue Curaçao—awful stuff, but perfect for making a Blue Hawaii cocktail.)

14 RIVERS

ÎLE DE MONTRÉAL, 1992–1996

Île de Montréal—which includes the city—is an island in the St. Lawrence River as well as a cultural and linguistic island. It has a larger population than Manhattan Island at just over two million. It is part of the Hochelaga archipelago, a cluster of islands at the confluence of the St. Lawrence and Ottawa rivers, which flow west inland to Ottawa and Lake Ontario, respectively.

I lived in Montreal for half of my twenties, in Little Portugal and Mile End, bumping into Leonard Cohen in the *dépanneur* on Rue Marie-Anne, drinking in Bar Miami and making pizzas at Euro Deli, and devouring the strange, cheap, beautifully broken, polyglot bohemian paradise that it was in the 1990s. (Unsurprisingly, this was the birthplace of *Vice* magazine in 1994.) There was an ongoing biker war, a major heroin problem, rampant government corruption, and an arson epidemic. You could smoke everywhere including banks and grocery stores, unofficially if not officially, and cigarettes were mostly bought and sold illegally. You could buy large bricks of discount cheese and cans of maple syrup from clandestine salesmen in bars that stayed open till 4am. Bathroom surfaces were covered in coke and people had sex wherever they liked. The stairs of all the houses were on the outside and twisted like corkscrews. Most of my friends from Vancouver Island including J moved to Montreal in their twenties, and some still live there.

"I have to keep coming back to Montreal to renew my neurotic affiliations," Cohen says on the jacket copy of his second collection of poems, *The Spice-Box of Earth* (1961). He often traveled to other islands, whether it was Manhattan or Cuba or Hydra in Greece. I didn't make it as far as Greece but I often rode the Haitian community bus from Montreal to New York, everyone an islander of some kind, where they served warm soda out of plastic bottles and spent the eight-hour journey in one big animated conversation. Like Cohen, I studied English at McGill, but I didn't write much poetry, and I doubt I had nearly his luck in love. As always, there are islands within islands: Cohen described the affluent Anglophone neighborhood of Westmount, where he grew up, as "a collection of large stone houses and lush trees arranged on the top of the mountain especially to humiliate the underprivileged."[1] Westmount and McGill were Anglophone islands in Montreal, just as Montreal is a cosmopolitan island in Quebec, and Quebec is a Francophone (metaphorical) island in Canada.

River islands are a unique category of islandness. They tend to be much closer to the mainland, since rivers are rarely as wide as oceans. River islands are also surrounded by fresh water, generally, which brings its own differences. (The St. Lawrence is a mix, or rather layers, of salt and fresh water for a lengthy stretch of river from the estuary at Île d'Orléans near Quebec City to Lac Saint-Pierre, just upstream from Montreal.) I remember taking the metro to the end of the line with my friend Mike, who loved to wander aimlessly in the style of the Situationist *dérive*, tracing the edges of the island, exploring its physical boundaries, and looking across to the smaller islands and the mainland. Along with Montreal, famous urban river islands include Chongming Island in the Yangtze River (part of Shanghai), and the elegantly named Salsette Island, home to Mumbai and a population of 20 million (though a bit of a cheat as it is bounded on one side by the Arabian Sea), as well as historic curiosities like Île de la Cité in the River Seine in Paris and Isola Tiberina in the Tiber in Rome. The UNESCO-protected old town of Lübeck in Germany, once the seat of the Hanseatic League (medieval ancestor to the EU), is on an island in the River Trave, though you would hardly know it to stand there. And then there is Manhattan, that great urban river island, about which Camus wrote on a visit: "Sometimes from beyond the skyscrapers . . . the cry of a tugboat finds you in your insomnia in the middle of the night, and you remember that this desert of iron and cement is an island."[2]

One of the strangest river islands I've encountered is D'Oyly Carte Island, in the River Thames in England, which was bought by the Victorian theater impresario Richard D'Oyly Carte in 1890—not long after he had enlisted Oscar Wilde to serve as a human advertisement for Gilbert and Sullivan's comic opera *Patience; or, Bunthorne's Bride* on a yearlong tour of America. D'Oyly Carte wanted to use the private island as an extension of his grand Savoy Hotel in London, just a short boat trip away, but when he failed to obtain an alcohol license he built a mansion there instead and lived in it off and on until his death in 1901, a few months after Queen Victoria herself. He is rumored to have kept a crocodile.

My father gave me a mass market paperback of Cohen's novel *Beautiful Losers* (1966) when I moved to Montreal, and I still have it on my bookshelf. The inscription simply reads: "*bon voyage!* –RH." Cohen wrote the book on Hydra, taking acid and speed and typing all day in the hot sun. The passage I remember most, because it is so true, comes near the end of the novel: "In Montreal spring is like an autopsy. . . . From the streets a sexual manifesto rises like an inflating tire, 'the winter has not killed us again!'"[3] The long and brutal Canadian winter partly explains the hedonism of spring—as well as Cohen's decision to live half the year on Hydra with Marianne Ihlen. But Montreal's islandness is also crucial to its identity. As in Madeira or Corsica or Scotland, the dream of independence is deeply held in the province of Quebec. The Island of Montreal, however, was blamed in the 1995 independence referendum—the second in Quebec history—for the "Yes" side losing by only 50,000 votes. (Voter turnout was an astounding 93 percent. I voted too.) In an ill-tempered, anti-Semitic, xenophobic, and possibly drunken concession speech, the leader of the independence movement, Jacques Parizeau, blamed the narrow loss on "l'argent pis des votes ethniques," ("money and the ethnic vote")—in other words, the cosmopolitan island of Montreal which sat in stark contrast to the "pure laine" ("pure wool") French-Canadian settler population dominant in the rest of Quebec. But Montreal wears its difference proudly: it is one of the most diverse places I have ever lived, where people often speak three languages at once and, like pre-gentrified Manhattan, its inhabitants are proud of the chaos and cacophony of its many distinct neighborhoods sitting side by side. I remember the morning after the referendum and Parizeau's bitter concession speech: people around me were saying that if Quebec deserved independence from Canada, Montreal deserved to be a sovereign island within Quebec. (This message goes into a corked bottle of La Fin du Monde, from Montreal brewery Unibroue.)

15 IMAGINARIES

COUGAR ISLAND, 1977–1980

The things we don't know about our families would take another lifetime to learn. I saw the tip of this iceberg when my aunt told me the true story of her older brother's death in a plane crash on the Canadian prairies in 1964. She had only learned the truth herself by talking to her brother's friend, now in his seventies, who lives on Salt Spring Island near Vancouver Island. She was only thirteen when he died, and my father was seventeen. The brother was flying the plane and another man was taking aerial photographs. Both died in the crash. The story I always heard, the official story, was that it was a simple tragedy: probably engine failure or inexperience. In my grandfather's memoir the brief mention of the event describes the two men being "killed while taking low level pictures of farmsteads from a Luscombe aircraft." Now my aunt told me she had learned from the brother's friend that both men were found to have a blood alcohol level above 0.2 percent, which for most people means severe intoxication, vomiting, even blackouts. That might explain the crash. The friend also told her that her brother was gay. This was something else she had not previously known, and something it wasn't easy to be in early-1960s Alberta, even coming from a socially liberal family. I wondered if my father knew; she wasn't sure. Suddenly I felt closer to the uncle who died almost a decade before I was born. I wished I had known him—that he had made it to the island with the rest of the family.

FIGURE 10 Father and Son, 1970s.

I asked my mother how we all came to live on Vancouver Island. My relatives on both sides were from the prairies, although they moved constantly and all of their ancestors came from islands. It took her a minute to get it straight. "It was my parents who arrived first, because it was cheap and warm," she said. "Cheap?" I asked, in disbelief. "They wanted somewhere to retire when my dad got out of the air force. They found their house, the one you remember, on eleven acres near the old speedway. It was only $27,000. It would be worth a hundred times that now. I remember seeing pictures before we moved—it was so green, and the mountains! All I knew was flat and grey and cold." When she came to the island her first impression was the scale—almost miniature compared to the open prairie, and reassuringly bounded. She mentioned other reasons to live on an island: finding or forming a small community, being apart from the mainstream, and being surrounded by the sea. We talked about a "three bears" approach to choosing an island: finding the right size, the right distance from the mainland, and the right degree of isolation. Vancouver Island was large, close, and—ferry waiting times aside—enjoyed all the modern conveniences, but it was also a calm, quiet, veritable utopia a world away from the rat race of mainland Vancouver.

Halfway through the history my mother realized she'd made a mistake—it was her brother who arrived first. "Angus was flying fighter jets out of Comox. They used to fly F-18s up there, now it's just search and rescue. So I guess my parents visited Angus and decided it was the place to retire. Bobby came with them because he was just finishing high school. He got married, joined the army, got divorced—all by twenty. Life moved fast back then, you'd be surprised." My mother had also married at 19, had me at 21, and was divorced by 25. "After we split up your dad moved back East for a teaching job, but he couldn't stand being away from you, so we both ended up moving to the island. His parents moved here a year or so after that, with your aunt. Your dad's brother died in the plane crash some years before that. Then a few cousins came, and your aunt Vi, and pretty soon it was the whole family."

Cougar Island was where my single dad used to take me camping every summer. There was a small car ferry, and on the crossing I always fell asleep, making it hard to judge the time and distance. We drove off the ferry and straight into the forest, and before long we would find a place to set up camp among the trees and quivering ferns. The earth was soft enough that you didn't need a mattress as long as you stayed dry. Staying dry was hard: our pup tent always sagged and leaked. For dinner in the darkness around the fire he cut Spam into slices with his pocketknife and fried them in a cast-iron pan with pork and beans. There were bears in the woods—only island black bears—so we had to hang the food in the trees or lock it in the car. They came and sniffed around anyway, smelling traces of food, while we huddled inside the tent listening to their snuffling with only a thin layer of nylon separating the bears and us. We would go for long walks through the impossibly tall, dark forest.

One time he lost me—I was only five or six. I wandered along a clear, narrow stream as dusk fell, crunching rocks underfoot. He called and called until eventually he found me, snatching me up in his arms, his flannel shirt damp with sweat and smelling of campfire smoke. I remember the feeling of safety as he pulled me up onto his shoulders and carried me back to camp.

In the documentary *Beaches of Agnès* (2008), the French director Agnès Varda says that if we are lucky we imagine our close family as a "peaceful island." Until the age of ten, when my father remarried (briefly), my parents were not so much a peaceful island as two islands in an archipelago. I navigated between them, a figure in a small boat. But soon I became an island too. It's a tendency I fight against in myself: the tendency towards isolation, feeling estranged from the mainland of humanity. My mother has expressed similar feelings: autonomy, self-sufficiency, just enough and no more. A stoic insularity born on islands thousands of miles away in the North Atlantic and passed down through generations. Though he could be sociable and very funny, the lasting image I have of my father is a solitary man, a hoarder in reverse, lying on the sofa in an empty house.

My dad made a lot of things up when I was a kid, which I found both fun and confusing. He planted spaghetti in the garden and made it grow an inch every time I came to see him, until it was tall enough and we harvested the dried pasta for dinner. He made up facts and became various characters. He had a lightly sadistic side, maybe something he learned from his older brother, or from missing his brother. He laughed when he told me about tying my aunt to a tree and leaving her in the yard all day. He would take me out for bike rides and lose me on purpose so I'd have to find my way home, claiming it helped my sense of direction. He was a loving parent, but as well as losing his older brother he had grown up with a father who had escaped the hardship and conflict of his own little island and did not suffer fools lightly.

I was probably sixteen when I finally learned that there was no Cougar Island: the island where we had camped every summer did not exist. It was a campground called Bamberton, just another part of the same island, reached by crossing Saanich Inlet from Brentwood Bay to Mill Bay. We were on an island, but it was the same island we were always on: the old island seen through new eyes. Because of the name it felt exciting, a dangerous place to visit. A thrill ran up my spine when I imagined how nervous my mother would get when he announced where he was taking me: "Not . . . *Cougar Island*?" (I'll put this message in a Molson stubby, the kind of beer my dad used to drink.)

CODA

A PACKET OF
LETTERS ON CRETE

I'm finishing this book on an island. Madeira was too expensive in summer, so we ended up on Crete during a record heat wave. As I write this, massive fires are burning on the Greek islands of Rhodes, Corfu, and Evia, as well as in Sicily and all across the mainland Mediterranean. Gran Canaria and Tenerife are burning in the Atlantic, and in the Pacific the island of Maui has been hit with catastrophic fires and is already being called a "warning" of disasters to come.[1] The temperature is 104 degrees Fahrenheit. Everything is bone dry; the plants look as though they have never seen rain. A few days ago a funeral was held for a firefighter from the island who was killed fighting a fire on the mainland. At the beach this afternoon we watched a plume of smoke appear in the distance, and then a line of fire trucks rushing to put it out. Everyone on the little beach stopped and watched.

A wall of cicada noise forms the soundscape to the heat of the day as I write. The mornings and evenings are significantly better: breakfasts of Cretan thyme honey in yogurt and thick black coffee; delicious green grapes that vary in size and don't taste of pesticides; late dinners of feta salad and other things you don't have to cook, and good local wine. I can't help but think of the Paul Mazursky film *Tempest* (1982), an eccentric adaptation that relocates the action to a modern Greek island. It has a wild cast, including John Cassavetes as the Greek-American architect going through a midlife crisis, Gena Rowlands as his estranged wife, Molly Ringwald as his teenage daughter, Susan Sarandon as his island lover, Raul Julia as Caliban (or "Kalibanos"), and Cookie Mueller as "New Year's Eve Party Girl." The film captures the arid otherworldliness of Greece and the odd motivations that carry people to islands to act out their personal dramas.

Despite the heat, it's good to be back on an island. Many things are better since we moved to the mainland, but it's different, and I understand the complicated relationships people have with islands they've left. As in Madeira, and most warm islands I have visited, the principal activities are swimming and eating. The people seem kind and relaxed. As Laura Jesson muses in the David Lean film *Brief Encounter* (1945): "I believe we should all behave quite differently if we lived in a warm, sunny climate all the time. We shouldn't be so withdrawn and shy and difficult." I've become one of those northern mainlanders who pines bitterly all winter for a ray of sunshine or a wave across my bare feet on warm sand.

Simone hunts for glass and pottery among the rocks and shells, and seems more at home here than on the mainland. I've become reacquainted with the swallows that swoop and dive at dawn and dusk, the lizards you catch out of the corner of your eye, and cockroaches that glide across the floor like wind-up toys. There are mice behind the fridge and a herd of goats living next door.

I promised myself once I got here I would finally open a packet of letters, the ones that were found by J's sister and given to me last winter by his first wife. To be honest, I would rather keep all that repressed—leave it on that other island where it always rains. But the trip is almost over, and the book contract too, so I'll break the seal and read my friend's messages in a bottle.

Opening the large manilla envelope, I am initially disappointed to see my own handwriting on most of the letters that fall across the desk. My juvenile reflection stares back at me: passing ideas about art and life, long forgotten. "I'm in a cafe, reading Foucault and thinking about Leila"— that kind of thing. I sift out my letters, flattered they were kept but not interested in reading them. What remains are a few scribbled and unsent letters from J, as well as odds and ends he meant to enclose, advertisements and labels from old bottles. There is a whole pad of writing and sketches, apparently directed to me but possibly also to himself. He was teaching himself to play the piano, he said (he eventually learned). He was making up abstract compositions but insisted they were only "a jumble of notes" because he didn't want to be like those fakers in art school. It is winter; he asks if it's snowing in Montreal. He's learning to sculpt the way he paints, transposing the ideas of one medium onto the other.

I pick up another letter, more formal in appearance. He is telling me about a dream. The two of us are walking down the middle of a road, past street after street of derelict buildings, in the dusty light he remembers from Africa, and it looks suspiciously like the apocalypse. Then we pass through a wall and we're in his elementary school. That's the whole letter, closed off with a messy signature. Returning to the first letter, written on the pad, I notice more pages of writing after a dozen or so sketches. It seems to be a diary rather than a letter, but it might be both. He is in London: walking from Portobello Road to Soho, becoming exhausted, falling asleep on a park bench, selling his rail pass to a man in the station and buying a pint and a meal with it. At the bottom of the last page there is a simple but elegant drawing of his studio. He signs off in a scribble: "This is my doctor's writing."

In 2019 I wrote a pitch for a short book called *Island*—so simple, as it seemed to me then; a book about islands and islandness, or islands in the digital age, or something. Then came the pandemic, followed by a war in mainland Europe—I revised the pitch in the summer of 2021. A few weeks later J was killed and everything shifted again, including the book. I'm reminded of the first part of *To the Lighthouse*, before the war, when Mrs. Ramsay listens to the sound of the waves crashing on Skye and thinks of the sound as a lullaby offering safety and support (as later Lily, the artist, thinks: "The sigh of all the seas breaking in measure round the isles soothed them").[2] But she also hears something more ambivalent, even ominous: "at other times . . . [the waves] had no such kindly meaning, but like a ghostly roll of drums remorselessly beat the measure of life, made one think of the destruction of the island and its engulfment in the sea, and warned her . . . that it was all ephemeral." She senses the terrifying passage of time, human mortality, even geologic time, and it fills her with terror—and she is right, for soon she is gone, along with the war dead.[3]

As Charmian Clift wrote from Hydra: "On an island, eventually, you are bound to meet yourself."[4] Just as people go to islands to discover themselves, I hoped to discover something about myself and islands by writing about a lifetime of island experiences. (Discovery is the task of the essayist: which is why, as I tell my students, relying too much on an AI assistant is counterproductive. What will you discover?) Island experiences are more than just a tan and a swim or an adventure; they offer us the chance to see life in a more essential, physically limited, form—to see our lives laid bare and to ask ourselves what of all this really matters. Is it family, or friends, or solitude? Art or experience? When we see our life stripped to its essence, what do we value most? Peter Conrad views islands as "existential terrain."[5] For Clift, islands are double-edged: they represent depression and elation, terrifying possibility, "the wide blue frightening loneliness of freedom."[6] (Clift, whose longing for oblivion is already hinted at in *Peel Me a Lotus*, died by suicide not long after returning to Australia with George Johnston and their children.) Something about islands makes them spaces to reflect and brood, away (as a visitor) from mainland concerns. Woolf considered *To the Lighthouse* an elegy: for her parents, as well as for everything lost in the war. She made the conscious choice to transfer the setting of the novel from St. Ives on the Cornish coast, where she spent her childhood summer vacations, to the Isle of Skye. When they finally reach the lighthouse, the children view their mythical, terrible, inscrutable father: "He sat and looked at the island and he might be thinking, We perished, each alone, or he might be thinking, I have reached it. I have found it, but he said nothing."[7]

I think about how J would have loved drawing this island, so different from the Pacific Northwest, and I picture him outside the house with a sketchpad on his knee. The letters reminded me what the real person was like, beyond the image I've kept of him since his death. Like my father, J was beloved and revered but he wasn't a saint—both men could be maddening in their disregard for others and their single-mindedness. J told me in Madeira shortly after his dramatic split with his first wife that he wanted to live like an artist. I didn't think that was fair, I said: he seemed to have that life already. But he led the life he wanted, selfishly or authentically or both, until his death. I'll put this message into the last bottle of Bushmills, which is now empty. My crate is full and ready for the sea.

I didn't read any poetry at J's funeral, but in retrospect I might have recited some famous lines by a poet he loved. I always found Yeats's "The Lake Isle of Innisfree" (1890) both poignant and silly, like most Yeats—the cabin "of clay and wattles made," the "bean-rows" and the "bee-loud glade." It occurred to me: what does the lake isle actually look like, and does it even exist? I Googled it from this rented cottage on Crete, and found a lough in County Sligo with a jetty at the end of an unnamed road. You can row, if you have a rowboat, the short distance to the uninhabited lake isle, which is truly tiny but looks quiet and idyllic. I will leave you there: with the sound of lapping water, in peaceful isolation. We might be islands, or perhaps death itself is an island: but if we're lucky we will be surrounded—or remembered—by those who love us. I hope J is at peace, and my father the lonely gardener as well. I miss them both terribly. Unlike my father, or Yeats, rather than live in romantic solitude I would choose to go to that lake isle with Simone: she could tend the bean-rows, while I try to figure out apiculture.

DESERT ISLAND DISCS

EIGHT SONGS ABOUT ISLANDS

1. Cesária Évora, "Isolada"
2. The Paragons, "The Tide is High"
3. U. S. Girls, "The Island Song"
4. Mona Baptiste, "Calypso Blues"
5. Autechre, "Montreal"
6. De La Soul, "Wonce Again Long Island"
7. The Pastels, "Leaving This Island"
8. Sergei Rachmaninoff, "Isle of the Dead (Opus 29)"

NOTES

Chapter 1

1 Anna Holligan, "Amsterdam launches stay away ad campaign targeting young British men," *BBC News*, March 29, 2023, accessed June 3, 2023, https://www.bbc.com/news/world-europe-65107405.

2 Hemingway kept many cats in both island homes. The descendants of Hemingway's six-toed cats, around sixty at present, continue to live and thrive in the house on Key West (https://www.hemingwayhome.com/our-cats).

3 Judith Schalansky, *Pocket Atlas of Remote Islands*, trans. Christine Lo (New York: Penguin, 2012), 23.

4 Katerina Teaiwa, "No more drinking water, little food: our island is a field of bones," *Guardian*, November 2, 2022, accessed June 5, 2023, https://www.theguardian.com/world/2022/nov/03/no-more-drinking-water-little-food-our-island-is-a-field-of-bones.

5 Tove Jansson, *The Summer Book*, trans. Thomas Teal (London: Sort Of Books, 2003), 27-8.

6 Charmian Clift, *Peel Me a Lotus* (London: Muswell Press, 2021), 89.

7 Peter Conrad, *Islands: A Trip Through Time and Space* (London: Thames & Hudson, 2009), 6.

8 Olga Tokarczuk, *Flights*, trans. Jennifer Croft (New York: Riverhead, 2018), 83.

9 Oliver Barnes, "No phone, no food—the wild rise of the survival holiday," *Financial Times*, September 7, 2023, accessed June 6, 2023, https://www.ft.com/content/e1b43290 -8425-4266-83b8-dc1b870e9049.

10 D. H. Lawrence, "The Man Who Loved Islands," in *The Man Who Loved Islands: Sixteen Stories by D. H. Lawrence*, ed. Frances Wilson (London: Riverrun, 2021).

11 Elsa Vulliamy, "British woman rescued from Atlantic waters after swimming after a cruise liner she thought her husband was on," *Independent*, March 28, 2016, accessed June 3, 2023, https://www.independent.co.uk/news/world/europe/british -woman-rescued-from-atlantic-waters-after-swimming-after -a-cruise-liner-she-thought-her-husband-was-on-a6956931 .html.

12 Charles Darwin, *Voyage of the Beagle* (London: Penguin, 1989), 269.

13 F. Scott Fitzgerald, *The Great Gatsby* (New York: Macmillan, 1992), 110.

14 Roland Barthes, *Roland Barthes by Roland Barthes*, trans. Richard Howard, (Berkeley: University of California Press, 1994), 92-3.

15 Claude Coste, "Roland Barthes's Visits to Greece," trans. Sam Ferguson, *Barthes Studies* 5 (2019): 15.

Chapter 2

1 Douglas Coupland, *City of Glass* (Vancouver: Douglas & McIntyre, 2009), 10.

2 Stuart Hall and Bill Schwarz, *Familiar Stranger: A Life Between Two Islands* (London: Penguin, 2018), 8–9.

3 Emily St. John Mandel, *Station Eleven* (London: Picador, 2014), 74.

4 The island punk scene is captured well in Paulina Ortlieb's documentary *Somewhere to Go: Punk Victoria* (2017), available on YouTube, https://www.youtube.com/watch?v=LDtKAqhnKUY.

5 "Number and Rate of Homicide Victims, by Census Metropolitan Areas," Statistics Canada, accessed August 11, 2023, https://www150.statcan.gc.ca/t1/tbl1/en/tv.action?pid=3510007101.

6 D. H. Lawrence, "The Ship of Death," in *Selected Poems*, ed. James Reeves (London: Heinemann, 1967), 82–3.

Chapter 3

1 Thomas Bernhard, *Concrete*, trans. David McLintock (New York: Vintage, 2010), 142.

Chapter 4

1   Italo Calvino, "The Adventure of a Poet," trans. William Weaver, in *Difficult Loves* (Boston: Mariner, 2017), 151.

2   David Marchese, "A Genius Cartoonist Believes Child's Play Is Anything But Frivolous," *New York Times*, September 2, 2022, accessed June 20, 2023, https://www.nytimes.com/interactive /2022/09/05/magazine/lynda-barry-interview.html.

Chapter 5

1   Fernando Pessoa, *The Book of Disquiet*, trans. Richard Zenith (London: Penguin, 2002), 148.

2   Georges Simenon, *The Mahé Circle*, trans. Siân Reynolds (London: Penguin, 2014), 46.

3   Ibid., 55.

4   Ibid., 21.

5   Jansson, *Summer Book*, 13.

Chapter 6

1   Darwin, *Voyage of the Beagle*, 41.

2   William Shakespeare, *The Tempest*, ed. Cedric Watts (London: Wordsworth Classics, 2004), 2.1.48–49.

3 Amílcar Cabral, "A Ilha" (my translation), reprinted in Pedro da Silveira, "Um poema (quase inédito) de Amílcar Cabral," *Seara Nova* 1550 (December 1974), 36–37.

4 Shoshana Zuboff, *The Age of Surveillance Capitalism* (New York: Public Affairs, 2018), 5.

5 Homer, *The Odyssey*, trans. Robert Fagles (London: Penguin, 2006), 214.

Chapter 7

1 John Steinbeck, *The Log from the Sea of Cortez* (London: Penguin, 2001), 179.

2 Virginia Woolf, *To the Lighthouse* (London: Penguin, 2000), 83.

3 Lord Byron, *Don Juan* (London: Penguin, 2004), 407.

4 The shocking event was captured in their 1995 film *K Foundation Burn a Million Quid*.

5 More information about the journey to Orkney and other speculative energy experiments described in this book—all of them led by James Auger—is available on the Reconstrained Design website: https://reconstrained.design/.

Chapter 8

1 Joseph Phelan, "Svalbard: The Arctic islands where we can see the future of global heating," *Guardian*, May 13, 2023, accessed June 23, 2023, https://www.theguardian.com/

environment/2023/may/13/svalbard-the-arctic-islands
-where-we-can-see-the-future-of-global-heating.

2 An earlier version of this chapter first appeared on the Crap
Futures blog: https://crapfutures.tumblr.com/.

Chapter 9

1 John Urry, *Offshoring* (Malden, MA: Polity, 2014).

2 Ray Bradbury, *Yestermorrow: Obvious Answers to Impossible
Futures* (Santa Barbara: Joshua Odell Editions, 1991), 155.

Chapter 10

1 Georges Perec, *W or The Memory of Childhood*, trans. David
Bellos (London: Vintage, 2011), 66.

2 Jason W. Moore, "Madeira, Sugar, and the Conquest of Nature
in the 'First' Sixteenth Century, Part II: From Regional Crisis
to Commodity Frontier, 1506-1530," *Review (Fernand Braudel
Center)* 33, no. 1 (2010): 1-24.

3 Charles Moore, *Margaret Thatcher: The Authorised Biography.
Volume 1: Not for Turning* (London: Allen Lane, 2013), 116.

4 Joe Brainard, *I Remember*, ed. Ron Padgett (New York: Granary
Books, 2001).

5 George Monteiro, "Ernest Hemingway in Madeira in 1954,"
Portuguese American Journal, October 3, 2014, accessed on
June 16, 2023, https://portuguese-american-journal.com/essay
-ernest-hemingway-in-madeira-in-1954-by-george-monteiro/.

6 Simenon, *Mahé Circle*, 113.

7 https://www.instagram.com/tar_island/

Chapter 11

1 Flann O'Brien, *At Swim-Two-Birds* (Normal, IL: Dalkey Archive, 2001), 12.

2 James Joyce, *Ulysses* (London: Everyman's Library, 1992), 52.

3 R. F. L. Hanna, *Five Pounds to Canada* (Victoria, BC: privately printed, 1983).

4 The fullest account I have read is Theo McMahon, "The Tragic Deaths in 1871 in County Monaghan of Emily and Mary Wilde, Half-Sisters of Oscar Wilde," *Clogher Record* 18, no. 1 (2003): 129–45.

Chapter 12

1 Lawrence Durrell, *Bitter Lemons of Cyprus* (London: Faber, 2021), 6.

2 Howard Eiland and Michael Jennings, *Walter Benjamin: A Critical Life* (Cambridge, MA: Belknap Press, 2016), 207–9.

3 Quoted in Eiland and Jennings, *A Critical Life*, 3.

4 Along with Benjamin's letters and a few essays (e.g., "Ibizan Sequence"), and Eiland and Jennings's masterful biography, one of the most engaging accounts of Benjamin's time in Ibiza is

Frédéric Pajak's *Uncertain Manifesto* (trans. Donald Nicholson-Smith; New York: New York Review of Books, 2019), which he also illustrated.

5 Walter Benjamin, *Walter Benjamin's Archive: Images, Texts, Signs*, trans. Esther Leslie (London: Verso, 2015), 189–92.

Chapter 14

1 Leonard Cohen, *The Favourite Game* (Toronto: McClelland & Stewart, 2011), 48.

2 Albert Camus, *American Journals*, trans. Hugh Levick (London: Abacus, 1990), 51.

3 Leonard Cohen, *Beautiful Losers* (Toronto: McClelland & Stewart, 1991), 245.

Coda

1 Adrienne LaFrance, "Hawaii is a Warning," *Atlantic*, August 10, 2023, https://www.theatlantic.com/ideas/archive/2023/08/hawaii-wildfires-warning-climate-change/674974/.

2 Woolf, *To the Lighthouse*, 155.

3 Woolf, *To the Lighthouse*, 20.

4 Clift, *Peel Me a Lotus*, 71.

5 Conrad, *Islands*, 8.

6 Clift, *Peel Me a Lotus*, 197.

7 Woolf, *To the Lighthouse*, 224.

ACKNOWLEDGMENTS

I want to thank James Auger, whom I've had the great pleasure to work and barbecue with side by side for the better part of a decade, and who led the island projects described in this book. The tales of our journeys to Eday and Svalbard first appeared on our blog, Crap Futures. I also want to thank Haaris Naqvi, Christopher Schaberg, Ian Bogost, Hali Han, and all the good people at Object Lessons and Bloomsbury for their support throughout this process and for creating such a brilliant series. I want to thank my always inspiring friend Eric Craven and other friends who appear in the book, named or unnamed (please forgive me). I should also give a nod to literary Twitter (RIP) and the editors of little magazines who first published some of the essays reworked for this book, including *3:AM*, *Numéro Cinq*, and *Minor Lit[s]* (special shout out to Tomoé Hill, who pointed me towards Charmian Clift). Finally I want to thank my family of islanders, near and far—and Lina, Thea, J's family, and J himself, for his friendship of 35 years.

INDEX